GIGI KAESER

ABOUT THE AUTHOR

ILAN STAVANS, a novelist and critic born in Mexico, is one of the most distinguished Hispanic voices to emerge in the last decade. He holds a Ph.D. from Columbia University and teaches at Amherst College. His books include *The Hispanic Condition, Bandido,* and *On Borrowed Words,* and he recently edited an anthology of Pablo Neruda's poetry. Stavans has contributed articles and reviews to *The Nation,* the *Washington Post,* the *Boston Globe,* the *Miami Herald, Newsday,* and *Bloomsbury Review,* among others. He lives with his wife and sons in Massachusetts.

ALSO BY ILAN STAVANS

FICTION

The One-Handed Pianist and
Other Stories

NONFICTION

The Inveterate Dreamer

On Borrowed Words

The Riddle of Cantinflas

Octavio Paz: A Meditation

Bandido

The Hispanic Condition

Art and Anger

¡Lotería! (with Teresa Villegas)

CARTOONS

Latino USA
(with Lalo López Alcaraz)

TRANSLATIONS

Sentimental Songs,
by Felipe Alfau

ANTHOLOGIES

The Scroll and the Cross

The Oxford Book of
Jewish Stories

Mutual Impressions

The Oxford Book of Latin
American Essays

Tropical Synagogues

Growing Up Latino
(with Harold Augenbraum)

Wáchale!

EDITIONS

The Poetry of Pablo Neruda

The Collected Stories of
Calvert Casey

Isaac Bashevis Singer:
Collected Stories (3 volumes)

GENERAL

The Essential Ilan Stavans

Ilan Stavans: Eight Conversations
(by Neal Sokol)

Spanglish

THE MAKING OF
A NEW AMERICAN LANGUAGE

Ilan Stavans

HARPER PERENNIAL

NEW YORK • LONDON • TORONTO • SYDNEY

A hardcover edition of this book was published in 2003 by Rayo,
an imprint of HarperCollins Publishers.

HarperCollins books may be purchased for educational, business, or sales
promotional use. For information, please e-mail the Special Markets Department
at SPsales@harpercollins.com.

FIRST RAYO PAPERBACK EDITION PUBLISHED 2004.

DESIGNED BY BETTY LEW

The Library of Congress has catalogued the hardcover edition as follows:
Stavans, Ilan.
Spanglish : the making of a new American language / Ilan Stavans.—1st ed.
p.   cm.
Includes bibliographical references.
ISBN 0-06-008775-7
1. Mexican Americans—Language.   2. English language—United States—
Foreign words and phrases—Spanish.   3. English language—
United States—Foreign elements—Spanish.   4. English language—
Variation—United States.   5. Spanish language—Influence on English.
6. Language in contact—United States.   7. Hispanic Americans—Language.
8. Bilingualism—United States.   9. United States—Languages.
10. Americanisms. I. Title.
PE3102.M4S63   2003   2003046579
422'.461—dc21

ISBN 0-06-008776-5 (pbk.)

HB 12.14.2020

To mis estudiantes del Special Topics en Amherst College,
and to Josh and Isaiah,
mis beloved hijos espanglishados.

Slang, n.,

The grunt of the human hog

(Pignoramus intolerabilis)

with an audible memory.

—Ambrose Bierce

Contents

Spanglish

Spanglish

Introduction

LA JERGA LOCA

¿Cómo empezó everything? How did I stumble upon it? Walking the streets of El Barrio in New York City, at least initially. Wandering around, as the Mexican expression puts it, con la oreja al vuelo, with ears wide open. Later on, of course, my appreciation for Spanglish evolved dramatically as I traveled around los Unaited Esteits. But at the beginning was New York. It always is, isn't it?

I had arrived in Manhattan in the mid-eighties. My first one-room apartment, which I shared with three roommates, was on Broadway and 122nd Street. The area was bustling with color: immigrants from the Americas, especially from the Dominican Republic, Mexico, El Salvador, and Colombia, intermingling with students from Columbia University, Barnard and Teacher's College, and with future ministers and rabbis from Union Theological Seminary and The Jewish Theological Seminary. The ethnic juxtaposition was exhilarating indeed. But sight wasn't everything. Sound was equally important. Color and noise went together, as I quickly learned.

I was enthralled by the clashing voices I encountered on a regular walk in the Upper West Side: English, Spanish, Yiddish, Hebrew . . . Those voices often changed as one oscillated to different areas of the city: Arabic, French, Polish, Russian, Swahili and scores of other tongues were added to the mix. What kind of symphony was I immersed in? Was this the sound of the entire universe or only of my neighborhood?

There was a newspaper stand on the corner of 110th and Broadway, next to a bagel bakery and a Korean grocery store. I regularly made my shopping in those blocks, so I regularly stopped to browse. Newspapers and magazines in English predominated in it, and Chinese and Israeli periodicals were also for sale. But the owner displayed the Spanish-language items with emphasis: *El Diario/La Prensa, Noticias del Mundo, Diario de las Américas, Cosmopolitan, Imagen* . . . As a Mexican native, I often bought one of them in the morning, "just to keep up with what's up," as I would tell my friends. But to keep up with these publications was also to invite your tongue for a bumpy ride. The grammar and syntax used in them was never fully "normal," e.g., it replicated, often unconsciously, English-language patterns. It was obvious that its authors and editors were americanos with a loose connection to la lengua de Borges. "Están contaminaos . . . ," a teacher of mine in the Department of Spanish at Columbia would tell me. "Pobrecitos . . . They've lost all sense of verbal propriety."

Or had they?

My favorite section to read in *El Diario/La Prensa*, already then the fastest-growing daily in New York, where I eventually was hired to be a columnist, was the hilarious classified section. "Conviértase en inversor del Citibank," claimed an ad. Another one would state: "Para casos de divorcio y child support, llame a su advocate personal al (888) 745-1515." And: "¡¡¡Alerta!!! Carpinteros y window professionals. Deben tener 10 años de experiencia y traer tools." Or, "Estación de TV local está buscando un editor de líneal creativo. Debe tener conocimiento del 'Grass Valley Group VPE Series 151'. En The Bronx. Venga en persona: (718) 601-0962." One morning I came across one that an-nounced pompously: "Hoy más que nunca, tiempo is money." And I stumbled upon another that read: "Apartments are selling like pan caliente and apartments de verdad."

Today I use the term *hilarious* in a reverent fashion. Over the years my admiration for Spanglish has grown exponentially, even though I'm perfectly conscious of its social and economic consequences. Only 14 percent of Latino students in the country graduate from college. The majority complain that the cultural obstacles along the way are innu-merable: the closely knit family dynamic, the need to help support their family, the refusal to move out from home in order to go to school . . . And language, naturally: for many of them proficiency in the English language is too high a barrier to overcome. English is the door to the American Dream. Not until one masters el inglés are the fruits of that dream attainable.

Spanglish is often described as the trap, la trampa Hispanics fall into on the road to assimilation—el obstáculo en el camino. Alas, the grow-ing lower class uses it, thus procrastinating the possibility of un futuro mejor, a better future. Still, I've learned to admire Spanglish over time. Yes, it is the tongue of the uneducated. Yes, it's a hodgepodge . . . But its creativity astonished me. In many ways, I see in it the beauties and achievements of jazz, a musical style that sprung up among African-Americans as a result of improvisation and lack of education. Eventually, though, it became a major force in America, a state of mind breaching out of the ghetto into the middle class and beyond. Will Spanglish fol-low a similar route?

Back then, as my early immigrant days unfolded, it was easier to denigrate it. Asked by a reporter in 1985 for his opinion on el espanglés, one of the other ways used to refer to the linguistic juxtaposition of south and north—some other categories are casteyanqui, inglañol, argot sajón, español bastardo, papiamento gringo, and caló pachuco—Octavio Paz, the Mexican author of *The Labyrinth of Solitude* (1950) and a recipient of the Nobel Prize for Literature, is said to have responded with a paradox: "ni es bueno ni es malo, sino abominable"—it is neither good nor bad but abominable. This wasn't an exceptional view: Paz was one of scores of intellectuals with a distaste for the bastard jargon, which, in his eyes, didn't have gravitas. Una lengua bastarda: illegitimate, even wrongful.

The common perception was that Spanglish was sheer verbal chaos—el habla de los bárbaros. As I browsed through the pages of Spanish-language periodicals, as I watched TV and listened to radio stations en español, this approach increasingly made me uncomfortable. There was something, un yo no sé qué, that was simply exquisite . . . Of course, it took me no time to recognize that standard English was the lingua franca of the middle and upper classes, but its domain was in question in the lower strata of the population. In that segment, I wasn't able to recognize the English I expected to hear: monolithic, homogenous, single-minded. Instead, I constantly awakened to a polyphonic reality.

Depending on the individual age, ethnicity, and educational background, a vast number of dispossessed nuyorquinos spoke a myriad of tongues, a sum of parts impossible to define. Indeed, the metropolis seemed to me a veritable Tower of Babel. And among Hispanics—the rubric *Latino* was only then emerging—this hullabaloo, this mishmash was all the more intense.

Mishmash is a Hebrew term that means fusion. In and of itself, the word Spanglish is that mixture: a collage, part Spanish, part English. I'm an etymological freak, always on the lookout for a lexicographic definition. I've sought for the inclusion and explanation of the word *Spanglish* in dictionaries.

Doctor Samuel Johnson, an idol of mine, once said that "dictionaries are like watches: the worst is better than none, and the best cannot be expected to go quite true." Perhaps that is the reason why so many ignore so common a verbal phenomenon: the best lexicon is never good enough, or the worst will build a fence around it not to let undignified terms irrupt on its pages. For shouldn't a lexicon, seeking to categorize the rowdy and infinite Spanglish exchanges, "to go as true as it might," in Johnson's mindset, begin by giving the meaning of the word?

Again, think of el jazz. In the seventies, Herbie Hancock offered a brilliant analogy: "It is something very hard to define," he said, "but very easy to recognize." Spanglish, I'm convinced, fits the same bill: it's not that it is impossible to define, but that people simply refuse to do it. And yet, nobody has the slightest doubt that it has arrived, que ya llegó . . . It is also a common vehicle of communication in places like Miami, Los Angeles, San Antonio, Houston, Albuquerque, Phoenix, Denver, and Tallahassee, as well as in countless rural areas, wherever the 35.3 million documented Latinos—this is the official number issued by the 2000 U.S. Census Bureau, which in 2003 jumped to 38.8 million, de facto making Hispanics the largest minority north of the Rio Grande—find themselves.

And, atención, Spanglish isn't only a phenomenon that takes place en los Unaited Esteits: in some shape or form, with English as a merciless global force, it is spoken—and broken: no es solamente hablado sino quebrado—all across the Hispanic world, from Buenos Aires to Bogotá, from Barcelona to Santo Domingo.

Beware: Se habla el espanglés everywhere these days!

To contradict Paz—and perhaps to correct him—let me attempt my own definition of Spanglish, once as succinct and encompassing as possible: "**Spanglish,** n. The verbal encounter between Anglo and Hispano civilizations."

As is always the case with these types of dictionary "explications," it already makes me unhappy. For one thing, I was tempted to write *clash* instead of *encounter*, and *language* instead of *civilization*. But then again, by doing so I would have reduced Spanglish to a purely linguistic phenomenon, which it isn't. Para nada . . .

At any rate, one thing is to get exposed to Spanish in the streets of New York, another altogether different is to use it effortlessly. As an immigrant, my road to full participation in American life was—as it has been and continues to be for any immigrant, regardless the origin—through English. I had come with primitive skills in Shakespeare's tongue, so during almost my entire first decade this side of the Rio Grande, my sole objective was to master it de la mejor manera posible, to the best of my capacity. Spanish was the language of the past for me, English the language of my future. It was only when I was already comfortable in both Spanish and English (as comfortable as one is ever likely to be) that I suddenly detected the possibilities of Spanglish.

This sequence of events, no doubt, has enlarged my overall appreciation of it. I date my full-grown descubrimiento in the early nineties. By then I had already left Manhattan and was living in a small New England town, where I taught at a small liberal arts college. My responsibilities included courses on colonial and present-day Latin America, and, on occasion, also on Hispanic culture in Anglolandia. The latter courses were invariably more challenging for me to teach. Students didn't register for them with the mere hope of learning about a specific period in history. Instead, their objective was psychological; they were eager to turn the classroom into a laboratory of identity.

They wanted to ask out loud: Quiénes somos? What makes us unique? And why are we here? Are we members of a single minority—the Latinos—or are we instead peoples of different ethnic, class, religious, and national backgrounds? In the isolated milieu of Amherst, Massachusetts, the language my Latino students used was recognizable to me. But I didn't pay much notice until Lisa Martínez showed up. (The name is fictitious and so are some of her circumstances.) Or better, until she was a punto de desvanecerse, about to disappear.

Originally from Istlos (e.g., East Los Angeles), Lisa, a junior, had taken a number of classes with me: on popular culture—comic strips, TV soaps, thrillers, music . . . —on autobiography, and on Argentine letters. We had established a solid relationship. Her odyssey was remarkable: Lisa had grown up in the inner city; she had been an active gang member and had seen a number of relatives and friends shot or imprisoned—vapuleados por el sistema; and she was initiated into Catholic life by an activist priest. Her tenure in Amherst, Massachusetts, was, hence, a radical change of scene for Lisa.

During her freshman year, Lisa felt disoriented, nostalgic for la casa, anxious to finish and return home. She also expressed her ambivalence at being an affirmative action student, enticed to the place by a full fellowship, but often looked at suspiciously by her Anglo counterparts because of her skin color, su pigmentación mestiza and her ethnic idiosyncrasy. Still, in her third year of college she appeared to have found inner and outer balance.

However, in recent times, whenever we stumbled across one another in the hallway, Lisa looked at peace with herself. It was somewhat surprising, therefore, that one frigid February, Lisa came to my office to say adiós.

"Ya me voy, profe . . . ," she announced.

I wasn't completely sure I understood her statement, so I asked Lisa where she was going. She answered that she was going back to her hood, to Califas, where people "no son tan fregados. They are más calientes, with a little bit of dignidad." Lisa was tired of the WASPy culture of the small liberal arts college she was invited to attend with an ethnic scholarship.

"Aquí no soy más que un prieto, profe. They want me pa'las quotas, so the place might say 'Chicanos are also part of our diverse student population.' Pero pa'qué, profe? I don't feel bien. I'm just a strange animal brought in a cage to be displayed pa'que los gringos no sientan culpa."

I begged her to be patient.

She was almost finished with her education, I said. Estaba casi de salida . . . One and a half more years—is that too much? But she wouldn't listen. Our entire tête-à-tête took approximately five minutes.

I never succeeded in changing Lisa's mind, nunca la convencí, and to this day I regret it. Somehow, seeing her walk toward the podium during Commencement to get her Bachelor's diploma would have been a better conclusion to the New England chapter of her journey.

In retrospect, Lisa's goodbye, su partida, was quite painful. Me rompió el corazón. I felt genuine affection for her. But the scene that took place between us in my office was more than about dropping out from college—at least for me. For, as I recall the occasion, the moment I opened my mouth, I realized every one of the words I uttered felt artificial, anomalous. I had wanted to tell Lisa that the separation from home is painful for everyone, that for some, like her, the separation isn't only emotional but also geographical and cultural. I told her it was important to keep in mind that H*O*M*E—and I pronounced the word patiently, comfortably, sweetly—acquires a different value, it becomes symbolic, the moment one leaves it behind. Or doesn't it? Even if she went back, her status as her mother's daughter would be different. And . . .

Pero no había vuelta de hoja.

To my consternation, though, I couldn't express myself. The more I tried to articulate my words, first in Spanish, then in English, the more dissatisfied I became. And why? Because I was overwhelmed with envy. To announce her sudden farewell, Lisa wasn't using the traditional college language a pupil is expected to articulate in the professor's designated space. Instead, judging by her vocabulary and syntax, she had already departed New England for Los Angeles: she was inhabiting the language of her turf, su propia habla, not the language of the alien environment where she found herself at present.

And what was expected of me: to ask her, in that troublesome situation, to switch to a more proper lengua?

That, no doubt, would have been counterproductive. What I desperately needed was for her to feel cómoda.

But what actually happened to me is that, instead of wanting her to "talk like me," my secret desire was the other way around: I wanted to use her own lingo.

Yes, en Spanglish—Lisa and I began to communicate in the jargon I had frequently heard, and had been enthralled by, en las calles de Nuyol.

Was I happy with my switch?

To my chagrin, I was . . . And what did I do? Nothing, absolutamente nada. I just let myself be taken by the verbal cadence of the conversation. Where is it written that faculty should elevate itself intellectually far beyond the reach of the students? Where does it say that professors cannot talk slang too? No sooner did I switch to Spanglish, though, that I realized that, as a teacher, I had crossed a dangerous line—una línea peligrosa.

My immediate, mechanical reaction was in tune with the milieu I came from as a middle-class Jew from Mexico whose choice it was to emigrate to El Norte: What on earth es ésto? I asked myself. Why was I mimicking Lisa? Wasn't my role as an intellectual and teacher to protect the purity and sanctity of el español and el inglés, rather than endorse this verbal promiscuity?

If I, and others like me, endorse this chaos of words, where is this syntactical amalgam leading us if not to hell? These were not easy questions. My tongue was moving in one direction and my heart in another. The more I rationalized what I was doing, the more guilt I felt. But I also realized that, through my standard English and Spanish, I lived in a verbal stratosphere remote from the universe I purported to invoke—

and teach—in the classroom, for pupils just like my dear Lisa Martínez. Besides, in Spanglish I felt freer, más libre. I didn't sense it as an imported, unnatural self. On the contrary, using it made me blissful.

Me sentí feliz!

Over the years, I've returned to that fated encounter hundreds of times. A door closed for Lisa Martínez that day but another one opened for me, a door se abrió and I walked through only to be radically transformed by the path I followed. That door has led me to more difficult questions.

*E*arly in the next academic semester after Lisa departed, I confided her story to a group of Latino students from different ethnic and geographical backgrounds. My remarks were focused less on Lisa's academic journey than on her way of communication. Piqued by my oddity, the students suggested the development of an independent study course on Spanglish. I responded enthusiastically. A full thirteen weeks to explore the myriad verbal possibilities—sin hesitación, without doubt, I realized that the experience would be rewarding in every sense. And it was, afortunadamante.

There are several ways to say *to reward* in Spanglish. One of theme is *reguardear*. And reguardados we became. The students, about 12–15 of them, and I, met regularly. Our first responsibility was to locate every possible item ever published on Spanglish. We were faced with a daunting reality: little was available on the subject. The frustration we felt was reminiscent of the one I came across at Columbia. In the classroom, the topics of discussion were Iberian luminaries like Lope de Vega, Quevedo, and Góngora; and also from Spain but closer to us, Benito Pérez Galdós and Leopoldo Alas "Clarín." From the Americas, the curriculum stirred us in the direction of Modernistas like José Martí and Rubén Darío. There was also the possibility of delving into "contemporary" figures, such as Jorge Luis Borges, Octavio Paz, Gabriel García Márquez, Mario Vargas Llosa, and the like. Their oeuvre was scrutinized in detail, and so was their Spanish. Conversely, the culture of Dominicans and Nuyorricans on the streets of New York—in fact, at the doorsteps of the building that housed us, La Casa Hispánica, on Broadway and 116th Street—was anathema: illegitimate, inappropriate . . .

Such negligence, me repetí a mí mismo, needed to be resolved without delay.

Lisa had invited me to revisit my overall identity as an intellectual of Mexican descent in the United States, as a professor, even as a father. Indeed, I began to rethink all my roles. My first child, Joshua, was only then two years old. Early on, I had decided to communicate with him solely in Spanish. Was Spanglish also a possibility now? Would he get confused if I used the slang, eventually being incapable of discerning the boundaries between languages? And should I talk to other students of mine in Spanglish the way I had done with Lisa? Could I eventually teach a course on the subject and write about it?

In those early days, a colleague of mine, aware of my interest, passed on a copy of a story, "Pollito Chicken," by the Puerto Rican author Ana Lydia Vega. It was part of a slim volume called *Vírgenes y mártires* (1981), but it was the sole piece in it not written in proper Spanish. In fact, it was, to the best of my knowledge then, the first full-fledged Spanglish story. So hypnotized was I by the experiment, I literally memorized—me aprendí de memoria—its three early sentences right away: "Lo que la decidió fue el breathtaking poster de Fomento que vio en la travel agency del lobby de su building. El breathtaking poster mentado representaba una pareja de beautiful people holding hands en el funicular del Hotel Conquistador. Los beautiful people se veían tan deliriously happy y el mar tan strikingly blue y la puesta del sol—no olvidemos la puesta de sol a la Winston-tastes-good—la puesta de sol tan shocking pink en la distancia que Suzie Bermiúdez . . ."

Without a doubt, Vega was ridiculing a generation of Nuyorricans ashamed of their island. But to me, the story carried a different message: it was less a political statement about the manufactured dreams of the Puerto Rican diaspora in the United States than an exploration of a slang that defines that diaspora. Eventually, I've learned that Vega has renounced "Pollito Chicken." When she published it, she was accused of chauvinism. Too bad, for the piece announces, in my view at least, a consciousness that would ultimately prevail, not only among Nuyorri-cans but also within the island itself: the Spanglish "I."

I was utterly mesmerized by "Pollito Chicken" and read it en público to my students. Their reaction was symptomatic: the Puerto Rican ones felt ambivalent toward it whereas those from the Bronx and Manhattan sympathized with its premise and celebrated its author as una original. That much was expected. But I was stunned by the fact that the Chicano, Dominican, and Cuban students were uninterested in it. There were portions of it they also didn't understand. The explana-tion, everyone in the group came to realize, was crystal clear. After all, the term Spanglish per se was an abstraction. For some Latinos, it is even offensive, denoting a broken frame of mind. Many prefer more at-omized rubrics: Cubonics, Dominicanish, Chicano Spanish, as well as Tex-Mex, Pachuco y un largo etcétera.

❧

*T*his dichotomy between the universal and the particular irrigates His-panic culture en los Unaited Esteits. Hispanic, a term that came about in the Nixon Administration, and its counterpart, Latino, are Platonic words that, para bien o para mal, symbolize a sum of parts. People from various national groups prefer to define themselves through more par-ticular names: Colombian-Americans, Ecuadorian-Americans, and so on. Therefore, the maxim *e pluribus unum* exemplifies the twisted dynamics within the community: the unity and the multiplicity are often at odds with one another, although for political purposes they often choose to create an alliance.

Hence, it quickly became obvious that there is really not one Spanglish but many . . .

The Spanglish spoken by Mexican-Americans in Istlos has its own characteristics that differentiate it from the Spanglish spoken by Cubans in Calle Ocho, on the other side of the country. To prove the point, that first semester we made an experiment. During family weekend, we invited to sit in the same room and socialize five Spanglish non-student speakers (parents, brothers, girlfriends) from different geographical parts of the country: a Cuban-American from Miami, a Mexican-American from San Antonio, a Nuyorrican from the Bronx, a Dominican-American from Washington, DC, and an Ecuadorian-American from Chicago. The sole mandate, la única y suprema obligación, was that they communicate with the rest in their own Spanglish. It quickly became clear that to be understood a number of terms, especially patronymics although not always, needed to be defined.

For instance, the Cuban-American student repeatedly referred to La Sagüesera, the southwestern sección de Miami. Nobody knew what she meant until she physically described the region. Likewise, el chicano from San Antonio talked about a washatería his mother owned and where he worked in the summer months. Only when another participant in the experiment asked if washatería was a laundry store did the rest know what the speaker was talking about.

Soon nos dimos cuenta that some of the participants employed only a handful of so-called "borrowed" terms, palabras prestadas, adapted—revamped, really—to somehow fit a Spanglish mode of communication, una manera de ser espanglishada. Others, instead, indulged in extensive code switching, el cambio de código, which is the way the easy transit between the languages is described by specialists in the field.

The tension between the one and the many is understandable. One is able to see it frequently on TV. But the moment one does, the drive toward a standardized form is established. It is a matter of letting others in on our individualized code. The tools to decode it are already in our culture, so it takes little time for people to understand one another. Indeed, as I began to research the fields of Spanglish, I realized that the bibliography on code-switching practices in individual regions (New York, Chicago, Colorado, New Mexico, Arizona and California) is extensive. But less substantial, nonetheless, are the studies on Spanglish modalities in the media. It is in the media, though, where Spanglish travels faster and the creation of a "common ground" becomes tangible. Univisión and Telemundo are the fastest-growing television networks in the United States. *El Show de Cristina*, *Sábado Gigante*, and *Noticiero Univisión*, to name only three of the most popular programs—and to obviate the ubiquitous soap operas to be found on prime time every weekday—are watched by millions.

The first two *programas* depend on guests. Those guests are average people invited to talk about their own life. Their expressions are full of *spanglishismos*. Every time one of these is repeated, the potential impact of the word is enormous. Terms like *parquear, grincar,* and *la migra,* which stand for to park, green card, and the staff of the Immigration and Naturalization Service respectively, have already become part of the lore.

Add to this the impact of radio. It is a well-known fact that there are more Spanish-language radio stations in the state of California alone than in all of Central America together. El impacto, pues, es asombroso. Almost without exception, these stations program call-in shows in which the listeners are able to express themselves on the air. Those that work as agricultural labor in the cotton, orange, and strawberry fields use these programs to let their relatives know where the next harvest is likely to take them. In urban centers, rap, rock, salsa and corridos in Spanglish spread the message out. The lingo de la calle y la montaña, then, penetrates people's minds, and their vocabulary, at an astonishing speed. La revolución lingüística es imparable—the verbal transformation is unstoppable.

One of my students showed up one morning with Spanglish versions of important documents. She started with the *Pledge of Allegiance:* "Yo plegio alianza a la bandera de los Unaited Esteits de América . . ." Everyone was laughing. That same student then moved on to a recitation of *La declaración of Independence:* "Nosotros joldeamos que estas truths son self-evidentes, que todos los hombres son creados equally, que están endawdeados por su Creador con certain derechos unalienables, que entre these están la vida, la libertad, y la persura de la felicidad."

And, to crown it all, she ended with *la constitución gringa:* "We la gente de los Unaited Esteits, pa'formar una unión más perfecta, establisheamos la justicia, aseguramos tranquilidá doméstica, provideamos pa'la defensa común, promovemos el welfér, y aseguramos el blessin de la libertad de nosotros mismos y nuestra posterity, ordenando y establisheando esta Constitución de los Unaited Esteits de América."

In my eyes, this was an exercise in ingenuity. It also showed astuteness, a stunning capacity to adapt, and an imaginative aspect to it that refuses to accept anything as foreign.

What was she really doing? The answer es sencilla: she was reappropriating major cultural artifacts that affected her life and that of everyone else. As a Nuyorrican, the student often talked of alienation. Se sentía ajena al quehacer nacional.

Indeed, for her the United States was a foreign country where her family of jíbaros had immigrated in the fifties as a result of the economic difficulties in Puerto Rico. The laughter in her face when she recited in Spanglish the information in the documents was far more harmonious than any other I had seen in her. "She's feeling comfortable," I told myself. "La *Pledge of Allegiance* and la *Declaration of Independence* son hers when she approached them through her own verbal prism . . ."

Livin' la jerga loca is how she described her effort . . .

An atmosphere of exhilaration invaded the classroom. All of us had suddenly become aware of the creative possibilities before us. I myself reacted by composing a gallery of Spanglish versions from famous first lines in American literature. For example, these lines from *Hojas de Grass* by the foundaing padre, Walt Whitman: "Sudenmente fuera del air estéril y drowsy, el lair de los esclavos Como un lightning Europa dió un paso pa'lante . . ." And the start of *Aventuras de Huckleberry Finn* by Mark Twain: "You no sabe de mí sin you leer un book by the nombre of *The Aventuras of Tom Sawyer*, pero eso ain't no matter." Or the one I like most: Robert Frost: *El Gift Derecho:* "La tierra was ours antes que nosotros were de la tierra. It was nuestra tierra más de cien años pa'trás . . ."

I remember perfectly the discussion that followed among us. Some students talked about understanding Spanglish through a generation lens. "My parents would hate these kinds of translations," one of them said. The rest of his argument was along these lines: "At home, this kind of stuff, like, it's really forbidden . . . Always, none of us would dare to use it at home, never. My father, in particular, he's a stubborn Cuban, an exile. He left the island in 1961. His entire life is devoted to the enlightenment of his children. He believes Cubonics to be a disgrace. Like, he hates it, he detests it completely. And, like, I tend to believe he is right. He is, because what would happen if all us used it? It would be terrible, wouldn't it? Spanglish isn't even a fully formed language. It is used by the common people, la prole, people without education, gente iletrada . . . My Pop's dream was to improve on our family condition. And language, I guess, is a lot about our own self, isn't it? Like, what you speak is what you are."

Another participant in the course was more forgiving. Her ancestors were from Venezuela but she had been raised in Madison, Wisconsin. She listened to raperos en español such as *Latin Alianza, Chicano 2 Da Bone, Latin Lingo,* and *Dr. Loco's Rockin' Jalapeño Band.* "These are the sort of lyrics these dudes include all the time. And it's cool! The songs make a connection. They express what the musicians feel . . ." She then showed the rest a CD she had brought along that day of a rapper group,

Ganga Spanglish. And in one the following sessions, she brought the lyrics of *KMX Assault.* I asked her to make me a copy. I remember these sentences: "Echar Pa'lante with my people is my imperative/ This Boricua will; endeavor to be clever." But her argument was that the slang would reach beyond the audiences that usually listened to those groups. "It's bigger than that," she said. "These roqueros are only reflecting what's happenin' en la calle . . . They aren't inventing it! Yeah, they are artists. But people wouldn't like their music if their songs didn't touch a cord. And they do! I know plenty of kids that listen to them. They memorize them. It ain't matter if you're puertorriqueña or mexicana. You listen to it because it's hip. Hip and hot!"

*T*hese opposing sides, as might be already deduced, were cut across class and ideological lines. Issues like bilingual education, affirmative action, and the impact of the English Only and English First movements determined people's views. One fraction believed that Spanglish was an obstacle for Latinos on the road to assimilation. Many of them neither speak Spanish nor English properly. Spanglish is an involuntary middle ground, from which they, if only they knew better, would like to escape pronto. Pitifully, musicians such as those of *Ganga Spanglish* exploited this limitation.

The other faction believed that Spanglish was a positive manifestation of the Hispanic spirit, that to speak a "broken" language was, in the academic lingo, a construction. By definition, the lower class is always less educated than the middle and upper classes. And it is left to those above it to ridicule its speech. In the end, though, it is the lower class where the most spontaneous aspects of culture are to be found. Sooner or later, others steal away those aspects, turning them into highbrow items.

In my view, the tension between these polar opposites was as important as the argument they set forth. I remember asking myself: Why is Spanglish so controversial? Why does it animate people as passionately as it does? The answer, I assume, is historical. Earlier in this essay, in my definition of Spanglish, I shied away from using the term *clash*. I guess my own political approach is made clear by the alternative: *encounter.* English and Spanish have found each other, they have become partners in this ever-expanding mode of communication. But that partnership—if that is what it is—has not always been around. Think of the defeat of the Spanish Armada Invencible by the British forces in 1588 and the embarrassment it brought along to the citizens of the Iberian Peninsula. King Philip II sought to undermine the Dutch by invading England. A huge patriotic campaign was orchestrated in Spain to justify the endeavor, accusing the Queen of England of heresy and pushing the concept of a Holy War as an excuse. But the enterprise was a disaster and an euphoric Spain was brought to its knees. Historian J. H. Elliott once argued that the "material effects" of the defeat were not striking: as many as two-thirds of a fleet of 130 ships returned home. Still, the psychological blow was unavailable. Not only was the Invincible Armada ridiculed, but the Iberian spirit began a long process of descent and disillusion. So yes, England, in the Spanish imagination, is el diablo—the devil. It brought down to its knees a commanding empire devoured by tax evasion and a lingering feudalism—then on its way out in most of Europe—and announced widely its ultimate demise. To this day, the country doesn't seem to have recovered in full. When more than twenty years later, the first part of *Don Quixote of La Mancha* was published, the national pride was already at a record low in spite of the battles against the Turks, which enabled Miguel de Cervantes Saavedra to explore the collective psyche through the dichotomy of reality and illusion. His errant knight seems to ask: Is Spain ever awake or does it live in a perpetual dream?

The clash between the Anglo and Hispanic habitats didn't stop there, though. The year 1898 marks the decisive Spanish-American

War in which Madrid was forced to depart from Cuba, Puerto Rico, and the Philippines, allowing the United States to move in as the new imperial orquestador of the region's affairs. This was another blow to Spanish self-esteem. It didn't come from London but from one of its former colonies, which was even more humiliating. It is not remarkable, then, that more than a century later, people in the Iberian Peninsula would approach, as they often do, the rapid growth of the Latino population in the United States as una forma de revancha—a revenge of sorts. It only takes a quick glance at the way news about it is reported to realize that the bruises of the past are still sensitive. The preeminence of Spanish as the second most important language in the United States encourages Madrid's government to move its strings as a form of support. "España está al centro del pasado y del presente de los Estados Unidos," King Juan Carlos de Borbón proudly announced in 1992, in honor of the Quincentennial of Christopher Columbus' descubrimiento of the Americas.

That same year, Puerto Rico, in a nationwide referendum, established el español as the island's official language. As a result the Spanish monarchy awarded all the Puerto Rican people the prestigious Príncipe de Asturias Prize. It was a political move, one meant to ratify the dominance of the Iberian tongue on the other side of the Atlantic.

Among educators and intellectuals, in diplomatic circles, for editors and reporters in the media, the presence of the Spanish language is an affirmation that the seeds of Spain's colonial quest are bearing fruit. Repeatedly, one reads about federal money in the Iberian Peninsula being allocated for programs to reinvigorate the maintenance and teaching of Cervantes's tongue this side of the Atlantic Ocean, and for institutions designed to foster the image of Spain among Hispanics.

Still, I prefer the term encounter, and not clash, simply because, as far as Latinos are concerned, these efforts by Spain are totally inconsequential.

How many Chicanos in the San Fernando Valley know of the mere existence of the Real Academia Española de la Lengua, an institution created in the 18th century to legislate—some would say promote—the well-being of the Spanish language? A minuscule number, no doubt. And how many Nuyorricans see their linguistic roots in Castile? An even smaller amount. . . .

Spaniards are known to be obsessed with language, but, after *Don Quixote,* they have not been particularly talented in producing a first-rate literature. And their language pride is colored by el remordimiento, abundant remorse. The thinker Miguel de Unamuno enjoyed making fun of those in his native Spain who claimed that people in the Iberian Peninsula in general need to learn their grammar and make sure that their stepchildren in the Americas use it as appropriately as they do. "The problem is not that Spaniards speak poorly," he said. "The real problem is that they don't have anything to say."

Nothing to say, eh?

Surely this wasn't the case with my students. The exhilarating experience with them made me fall in love with the labyrinthine nature of Spanglish. The more I reflected on it, the more I was mesmerized by its syncopated rhythms.

It was thanks to them that I understood that Spanglish cuts across economic terrain. It isn't spoken only por los pobres, the disposed. The middle class has embraced it as a chic form of speech, una manera moderna y divertida de hablar. This is in sharp contrast with other slang more often than not defined by turf: the language of drugs, for instance. Spanglish, instead, is democratic: de todos y para todos.

At any rate, by the late nineties I had begun to codify a lexicon of Spanglishisms. I'm not a linguist by training, nor am I a lexicographer by profession. In school I took a course on the history and morphosyntaxis of el español. My enthusiasm for words is that of an aficionado.

But Doctor Johnson was also an aficionado when he decided to codify his monumental *A Dictionary of the English Language,* wasn't he? And so was Sebastián de Covarrubias Orozco, officially "the first lexicographer" of Spanish, whose *Tesoro de la Lengua Castellana o Española* appeared in 1611, with the imprimatur of the Holy Office of the Inquisition.

Naïveté isn't always a handicap, and neither is amateurism . . .

I began to look around for any and all attempts to codify Spanglish. I found the important work done on the *variedades del español* in the United States. For instance, I realized that aside from a myriad of articles in books and scholarly journals by Samuel G. Armistead, John M. Lipski, Ernesto Barnach-Calbó, and Mary Ellen García, among others, Beatriz Varela had done a remarkable phonetic and morphosyntactical analysis of Cubonics in *El español cubano-americano* (1992). I also discovered the research by W. Labov in *A Study of Non-Standard English of Negro and Puerto Rican Speakers in New York City* (1968). Equally significant were the compilations by Rubén Cobos, a professor at the University of New Mexico, in *A Dictionary of New Mexico and Southern Colorado Spanish* (1983), as well as the lexicon by Roberto A. Galván and Richard V. Tescher, called *Diccionario del Español Chicano* (1989), and the one by José Sánchez-Boudy, under the name of *Diccionario Mayor de Cubanismos* (1999).

But these efforts were compartamentalized into national groups within the Hispanic minority. Instead, my interest was pan-Hispanic. I then invited my students to offer *palabras* they were accustomed to hearing at home, school, and in public spaces. Each made an initial harvest of almost a hundred, which increased as time went by. Over a period of several months, I was surprised to recognize an accumulation of approximately 1,000 terms.

The majority came from the United States but about 30% originated in places south of the Rio Grande, where Spanish and English were in contact.

My motivation increased . . . Every time I traveled to lecture, in the United States and abroad, I visited the local Hispanic neighborhood and talked to the people en la calle, en los restaurantes, en las canchas de soccer. Tape recorder and notebook in hand, I started to build upon a cross-cultural glossary. I also returned to 19th-century classics and spent time in libraries reading periodicals from the period published in New Mexico, Colorado, Arizona, California, New York, and Florida.

Over time, colleagues, friends, and acquaintances regularly e-mailed me additions and discussed variants and origins. Indeed, I realized that, as a result of American imperialism, Spanglish not only treks across economic lines but across national borders. This might be the byproduct of the Monroe Doctrine, which claims that América es para los americanos. What would he have thought of José Enrique Rodó, a leading figure of the Modernista movement at the turn of the 20th century, whose classic book *Ariel* (1900) in the form of a public epistle called for young Latin Americans to rebel against el dominio cultural norteamericano?

Surely, there is a difference between America and América: the former is an almighty, often bullish nation, the latter a romanticized continental landscape.

I came to realize that Spanglish wasn't, as is usually believed, a recent phenomenon. On the opposite, its past was fertile and far-reaching . . . My working definition of *Spanglish* became even more flexible: an encounter between cultures that is also a record of abundant past transactions. Think, by way of example, in the language of sports: *los doubles* in Tennis, *el corner* and *el ofsaid* in soccer, *el tuchdaun* in football, *el nocaut* in boxing. . . .

Or think of the business parlance: *marqueteo* and *la agencia de advertising,* for instance. And then there's Cyber-Spanglish, the cybernetic code used frequently by Internet users. The United States reached 128 million surfers in 2001. Countries like Mexico, Argentina, and Colombia have far fewer users, but their link to the World Wide Web, the so-called *La Red*, is also solid. Terms like *chatear, forwardear,* and *el maus* are indispensable north and south of the Rio Grande, as well as in Spain and in the Caribbean.

Or are they not? Is there a substitute in Spanish for each and every one of them? Probably there is, but it is absolutely cumbersome, which explains why, I guess, nadie lo usa, no one dares to use it.

In 1999 I officially taught a course called *The Sounds of Spanglish*. The central theme was the development of this mode of communication. It attracted a large student body. The key concept I used then, one that has stayed with me since, is *mestizaje.*

This palabra is still overcharged, está sobrecargada de historia. The Americas have been the site of cross-racial and cross-verbal fertilization ever since their entrance to modern times in 1492, if not before, as the aboriginal languages intermingled through war and domination across the continent. The arrival of the Spanish conquistadores and misioneros resulted in the rise of the mestizo, a by-product of East and West, of Iberian and pre-Columbian civilizations.

A racial category, the mestizo is a hand-me-down and also a half-and-half: his sight is divided between Europe and the archaic past. This duality was approached as a handicap for centuries. But as the Great Depression was sweeping los Unaited Esteits, the Mexican intellectual and once Minister of Education José Vasconcelos was among the first to champion the mestizo as a "brown race" in his book *La raza cósmica* (1929). He talked of mestizaje as a phenomenon of empowerment and announced that one day mestizos would be called to dominate the entire globe.

Vasconcelos's book is estrambótico, pamphleteering. His philosophical approach isn't rigorous enough. And his analysis of realpolitik leaves much to be desired. But this is not the place to question his arguments.

What interests me is the way those arguments have entered popular culture and how they have been assimilated by the Latin psyche? The student upheaval in Tlatelolco Square, Mexico, in 1968, and the Chicano Movement in the Civil Rights era, looked at Vasconcelos for inspiration. He was seen as an emblematic leader and his concept of mestizaje was used ideologically to advance the cause of anti-establishment forces. Through Vasconcelos's prism, the history of Latin America is defined by miscegenation. At the level of language, the Spanish spoken this side of the Atlantic Ocean also underwent a process of mestizaje, although less emphatically than the racial phenomenon.

No hubo un justo medio.

Think of the conquest of the Americas as the effort of one language to subjugate a plethora of others. For the effort of Spanish colonization not only gave room to a political, military, and social colonization of millions of people who belonged to a gamut of Indian tribes that spoke languages such Mayan, Huichol, and Tarascan in Mexico to Araucanian, Guaraní and Quechua in South America. It was, equally important, an act of linguistic enslavement, subyugación verbal. In 1492 Spain completed what has come to be known as La Reconquista, a project to make the kingdom fully Catholic and eliminate from it religious minorities, such as the Jews and Muslims. By then La Reconquista had lasted centuries—since the first crusaders of the 11th century. The result was that Castilian Spanish became known as the unifying tongue of the kingdom. That atmosphere gave way to a period of intense intellectual and artistic fertility. In the hundred and fifty years that followed, the so-called Siglo de Oro español, the Golden Age of Spanish arts and literature, mystical poets, playwrights and novelists, such as Fray Luis de León, Santa Teresa de Jesús, Lope de Vega, Francisco de Quevedo, Calderón de la Barca, Luis de Góngora, to name only the most prominent names, gave the world a taste of baroque sophistication, el saber barroco of Spanish courtly and country life.

The conquest of the Americas occurred at the same time. It was a massive undertaking, one that generated much controversy. Could the Spaniards "civilize" the native population? If so, what would be the toll? A debate in the so-called Nueva España, as Mexico was called in colonial times, between the Franciscans and the Jesuits, for instance, suggests there were opposing views at stake: one side believed the undertaking would result in "the Christianization of Mexico," whereas the other side was convinced the process would give room to "the Mexicanization of Christianity." Who was right?

Ambos, for transculturation was the outcome.

And did transculturation also occur at the level of language? Not quite . . . Given the overwhelming number of Indians, for a while, again in Mexico, Nahuatl was slated to become, in the eyes of the colonizers, the lingua franca of the region. But the thought was too threatening, demasiado peligroso.

Why Nahuatl and not el español? And did a jargon ever emerge in the Americas, one mixing pre-Columbian tongues and Spanish?

Obviamente no, since demographically it is unlikely: the Spaniards were few, the Indians were many but they were decimated by epidemics and malnutrition. For instance, chroniclers like "El Inca" Garcilaso in his *Royal Commentaries of the Incas* and Toribio de Motolinía in his *History of the Indians of New Spain* registered aboriginal terms. They were attuned to translation (un labio, dos labios) and recorded autochthonous terms as they heard them. The result is that today numerous Nahuatl words like *molcajete, aguacate,* and *huipil* (from *mulcazitl, ahuacatl,* and *huipilli,* respectively) have been accepted as *americanismos.*

This jumble, este orden en el desorden, suggests that the Americas exist in translation. That is, they are sensible to the imposition of language as a hegemonic force brought from abroad. By this I don't mean to suggest that when Sor Juana Inés de la Cruz and Augusto Roa Bastos wrote their poems and stories in Spanish, they first envisioned them in Nahuatl and Guaraní. Their language, Spanish, was—it still is and forever will be—as much theirs as it was the property of Benito Jerónimo Feijoo, Unamuno and Federico García Lorca. But the language as such wouldn't develop in the Americas the way it had evolved in the Iberian Peninsula: on this side of the ocean, it appeared suddenly and violently, spreading without mercy. Anything and everything that appeared before her was destroyed. Still, to this day it retains a slight degree of foreignness to it. In the Americas, Spanish is somewhat foreign, a sign of the imperial expansion of the Catholic Kingdom of monarchs Isabella and Ferdinand. El español como lengua extranjera. To show that this is the case, it is enough to look at the reaction to Spanglish on the two sides of the ocean: whereas Spaniards are often puritanical about their tongue, los americanos are altogether less hysterical about the issue as a whole. This, in my view, is because the population in Latin America is well acclimated to the act—y el arte—of colonization. They know by experience what it means to be subjugated by an alien tongue.

But, significantly, no middle ground emerged in Mesoamérica, no premodern Spanglish—not a mestizo Spanish but an in-between Spanish and indigenous tongues like Nahuatl.

It was not to be.

Amerianismos . . . Shouldn't the dictionary then define the terms originating, or at least in use, in the peninsula as iberismos? After all, the population of Latin America is approximately 350 million. Spain, in contrast, doesn't even reach 10 percent of that number. Una persona en Madrid might easily communicate with his counterpart in Caracas, but numerous nuances—from meaning to accent to emphasis—distinguish them. All in all, the americanismos are already the norm. Of course, there are many Spanishes south of the Rio Grande, as several more spoken north of it as well: Mexican, Cuban, Puerto Rican, et cetera, and within these categories there are different regionalisms, such as the immigrant Spanish from Nuevo León, Oaxaca, or Jalisco, for example; add to this the español novomexicano that is different from its many counterparts, such as the tejano, kanseco and californio Spanish. The result, clearly, is a Babelic jumble.

⊙⊚

*T*he extent to which *mestizo* Spanish and the pre-Columbian languages in the New World originally penetrated Iberian Spanish is illustrated by an anecdote about Antonio de Nebrija, the first to publish a grammar of the Spanish language, which appeared in 1492, the *annus mirabilis* in Iberian history. He included in his *Diccionario latino-español* the Latin term *barca* for a small rowboat. Then the *Vocabulario español-latino*, its companion volume, was released in 1495, and the Indian term *canoa*, from the Nahuatl, was listed, followed by the Latin definition. Evidently, in those three years, the impact of pre-Columbian languages on Iberian Spanish made itself felt. The Salamanca grammarian was only one of the conduits through which the verbal flux of Castilian Spanish began to manifest itself in the Americas. By devoting himself to standardizing and cataloguing the Castilian spelling and by studying its syntax and grammar, Nebrija legitimated a language whose speakers were only recently self-conscious of its global scope. Le dio a la lengua una presencia psicológica y nacional.

The vulgar Latin of the Roman Empire, which is different from the classical Latin of authors like Ovid and Seneca, gave room to a tongue— part of the family of romance languages that includes Italian and Romanian, influenced by Celtic and German, and by the Slavic varieties respectively—with a distinct flavor, su manera peculiar de ser. But Nebrija was only one among those responsible for the process of consolidation of Spanish as an hegemonic language.

The first official full-length dictionary of the Spanish language, the one by Covarrubias, was also a 17th-century by-product, albeit an indirect one, of the Universidad de Salamanca, where the compiler had been a student. (Among its alums, the institution also prides itself for having had Fray Luis de León, Calderón, and Unamuno parade through its hall-ways.) Very limited information is available about Covarrubias's academic qualifications. It is known that he was an ordained priest, a clerk and a religious instructor. Also eclipsed is the origin of his sole work, the *Tesoro*, which for years was referred to in various sources as *Etimologías* because of the emphasis it places on Latin, Greek and Hebrew etymological origins. By the way, Covarrubias also seems to have been versed, although less competently, in French and Italian, but apparently he knew nothing of Arabic, a major influence in the Spanish of the 10th to the 15th centuries.

Nebrija and Cobarrubias . . . I'm fascinated by their respective lexicons, their *joyas arqueológicas*, for they pose important questions as one ponders the status of Spanglish nowadays. Both the grammar by the former and the thesaurus by the latter were printed privately. They sold poorly. Nebrija was a committed scholar and his contribution to the legitimization of Spanish vis-à-vis Latin is unquestionable. Covarrubias's is another story. His *Tesoro*, as he argues in a note after the frontispiece, was made available so that Spain could match all the other nations that had already done the etymological work of releasing dictionaries. He hopes, as well, that not only the Spanish nation will find pleasure in it, "but also all the other nations that procure the ruin of our language with so much avariciousness." By royal decree, Italy and France established academies for the study and projection of the nation's language.

The *Accademia della Crusca* spent twenty years to prepare its own, which was published in six volumes in 1612; the *Academie Française*, whose mandate was "to purify" the tongue, worked on it from 1639 to 1694. Imitating them, Juan Manuel Fernández Pacheco, Marquis of Villena, founded the *Real Academia Española de la Lengua Castellana* in 1713, with Philip V's approval a year later. From its inception, the use of Spanish—or Castilian, as the institution ambiguously states even in its name—announced the dual desire to institutionalize one of the dialects of the peninsula and to safeguard it for posterity in its purest form.

Fernando Lázaro Carreter, a scholar and member of the academy since 1972, has explored in detail the shaping of the institution. In an extraordinary essay entitled *Crónica del "Diccionario de Autoridades" (1713–1740)*, he details the long struggle to complete a project that was really "larger than life."

Carreter explains how those that became the original members were, just like Covarrubias, neither lexicographers nor academics. They were devotees, based in Madrid, whose mission it was to replicate what the French had already done on the other side of the Pyreneans. The motto of the institution, much ridiculed in modern days, is "Limpia, fija y da esplendor"—cleans, standardizes, and grants splendor. The word *limpia* cannot but invoke the concept of pureza de sangre, purity of blood, through which the Spanish Inquisition propagated the idea that Old Christians made the nation proud whereas New Christians, e.g., crypto-Jews, also known as marranos, those that on the surface converted to Christianity before and after the expulsion of 1492 but remained de facto Jews at home, needed to be extirpated from the nation's landscape. In his mandate Philip II ordered that the academy publish a dictionary where words and their meanings would be normalized. The result was the famously disappointing *Diccionario de Autoridades*. It took from 1726 to 1740—veinticuatro años—to produce its six volumes. The dictionary sought to define a word and substantiate its definition with a quota of textual excerpts from established intellectual figures of the Golden Age.

In so far it was born as a replica, casi una copia Xerox, the dictionary of *Autoridades* was a disappointment. It was, no doubt, one of the most ambitious projects of the 18th century, but it was inundated by problems that, in many ways, need to be understood as symptoms of the Spanish character. For one thing, it served as a sideboard through which the institutional and less conscious censorship of the time manifested itself in it. The animosity against Jews, Muslims, and women, and the desire not to include rude terms and sexual innuendoes, are quite apparent in its pages. After the first volume had already appeared, the members of the *Real Academia Española de la Lengua* refused to introduce words—in Spanish they are called *voces*—that would be deemed offensive. Carreter himself explains the ups and downs of the volume's composition within the psychological and sociological context of the time. "The prestige of the work has not ceased to grow," he wisely states, "and today its esteem is unanimous."

> Founded on a few Spanish precedents, venerable yet quite imperfect in lexicographic terms, the Academy, established precisely with the objective to overcome those imperfections, devotes itself to the task of making an inventory, defining and authorizing with written texts, the fundamental mass of the Spanish vocabulary in only twenty-six years. That "only" has to do with the fact that the French Academy has taken sixty-five years in a much more limited enterprise. Six heft volumes, with a total of more than 4,000 pages, in quarto, were the result of that effort, one of the most aspiring that Spanish culture can be proud of. That *Diccionario,* from which more than two and a half years separate us, has not died yet: anybody interested in making a deep reading of a classic text still needs to consult it. And I don't exaggerate when I say, in many ways, it still has resolute strength in comparison with the most recent dictionaries, to which it is not unusual to compare it positively in its precision . . . I believe the simple chronicle of [its production] has some interest; it can be the mirror in which we can contemplate some aspects of the culture of that century, and a reflection, quite clear, of constant elements in our idiosyncrasy.

The dictionary of the Spanish language of the *Real Academia Española de la Lengua* used—and abused—presently, periodically updated in Madrid and nurtured with the input of the many branches the academia has in Latin America and the Philippines since 1871, is still based on the Covarrubias dictionary and on the *Autoridades*. It is somehow a legislative instrument that codifies and validates. Although other lexicons, algunas alternatives, might be purchased throughout the Hispanic world—including those by Corominas, Larousse, and Sopena, as well as my own favorite, by María Moliner, *Diccionario del uso del español*, released in 1966–67 and drastically revised and updated more than thirty years later, which stands as a towering achievement of individual Hispanic lexicography in a discipline known for its inexactitudes and ineficacias—it is this one that is endowed with an aura of astonishing power to accept or deny the legitimacy of any given word. Interestingly, in France the lexicon brought out by the *Academie Française* doesn't hold a similar place in the nation's culture; instead, the various dictionaries published privately, such as those by Hatier, Hachette and Robert, and especially Larousse, are the ones frequently sought.

España ha cambiado un poco, afortunadamente. Only during Franco's dictatorship, *especially* in its last stages, and more significantly with the arrival of democracy in 1974, has the Iberian Peninsula been ready to reflect on its linguistic heritage, and it has done so halfheartedly. The Socialist regime of Felipe González that brought not only social stability but an economic boom also announced an era of fractured autonomies, from Catalunya to the Basque Country, each with its ancestral tongue as a ticket of identity.

Increasingly, Catalan, Basque, and Galician are being recognized as separate languages in the Iberian Peninsula. Meanwhile, highbrow and popular animosity toward the *Real Academia Española de la Lengua* is as old as the institution itself. Accusations of elitism and pedantry abound. Frontal attacks in the press are always in store, and a series of parallel lexicons with terms grossly left out by the anointed erudite abound, of which *Diccionario de hispanoamericanismos no recogidos por la Real Academia* (1997), compiled by Renaud Richard, is only an example. It includes everything the dictionary produced by the royal erudite leave out—enough to have more than four hundred pages filled with diverse entries. Indeed, the accusations sometimes are humorous. For instance, the Cuban ethnographer Fernando Ortiz released *Un catauro de cubanismos* (1923), an engaging and encyclopedic attempt to codify the many idioms of Africans in Cuba. But this is also a volume displaying the varieties of Cuban Spanish not accepted by Madrid. Dictionaries are often arid and dumb in style. The one by Ortiz is filled with double entendres and an inimitable joie de vivre.

Herein one of its most incisive, critical entries:

> *Guayabo*—The tree that produces the guayaba, according to the *Diccionario de la Academia*. Why does it add: "In French: *goyavier*"? Does it mean to suggest that it is a gallicism? Really? Well, does the dictionary by any chance provide the French translation of every word? No? Then out with the *goyavier!* The etymology, if that is what is being proposed, is not worth a guayaba [no vale una guayaba], as we say. Let's call, instead, some of the twenty-two acceptions and derivatives of guayaba, cited by Suárez, that, like guayabal, guayabera, guayabito, would look better in the Castilian dictionary than that inexplicable Frenchified etymology. This guayaba is just too hard to swallow. (¡Que no nos venga la academia con guayabas!), and let us thus note, in passing, another Cubanism.

Ortiz accuses the Iberian académicos of elitism. Why look at France for etymological ratification, he wonders. For him a guayaba with a non-American taste is "too hard to swallow," a statement that encompasses the approach the *Real Academia Española de la Lengua*, and prior to it Covarrubias and Nebrija, have had toward the varieties of el español in the New World. Indeed, from Ortiz's statement one might deduce that the process of mestizaje—he preferred the term *transculturation*—is a painful one that the Iberian Peninsula is not able to fully recognize.

Nebrija once famously said: ". . . siempre la lengua fue compañera del imperio." Spanish was an imperial tool, indeed, with a clear-cut mission: to spread the faith of the Iberian knights and missionaries in uncivilized lands and force it onto the population. In the colonial period in the Americas, to civilize meant to reeducate, to evangelize, and to slowly incorporate the region and its inhabitants within the sphere of influence of the Catholic crown. The conquistadors and missionaries that arrived had the Bible in one hand and the sword in another. For them, language was an instrument to proselytize. But as ethno-linguist Angel Rosenblat has argued in his lucid *El español de España y el español de América* (1962), not even on this side of the Atlantic was it simply transplanted; instead, it adapted to the new reality by incorporating terms that came from pre-Columbian tongues, such as *aguacate* and *tenzontle*. Indeed, for over five hundred years Spanish has twisted and turned in a most spontaneous fashion from the Argentine Pampa to the rough roads of Tijuana. Today it is as elastic and polyphonic as ever, allowing for a wide gamut of voices that goes beyond mere localisms.

I was once asked by an Iberian reporter: Will Spanglish eventually replace Spanish? And will there ever be an *Academia del Spanglish?* My answer: let us focus on the present tense.

It is useful, for instance, to contemplate the passionate debate about it at the heart of the *Real Academia Española de la Lengua.* Its supporters include Carreter, whose views on the subject are lucidly expressed in the volume *El dardo en la palabra* (1997), a collection of newspaper columns on language, as well as a younger generation inducted to the academy and lead by the journalist Juan Luis Cebrián, whose seminal work as the first editor in chief of *El País* established him as a valuable intellectual voice of the post–Franco era. This debate is also palpable in the Hispanic intelligentsia at large. In 1997, the first Congress on the Future of the Spanish Language took place in Zacatecas, Mexico. Among those who participated were Gabriel García Márquez, Alvaro Mutis, and Camilo José Cela.

The author of *One Hundred Years of Solitude*, in his speech, traced the pattern a word used in Colombia takes to be incorporated into the *Diccionario de la Real Academia*. "The route takes an average of twenty years," he said, "and that is only if the word is accepted." He discussed the 107 synonyms for *penis* found in Quito, Ecuador—such as *churuca, micho, palitroque,* and *zirindango,* all recorded in *Léxico sexual ecuatoriano y latinoamericano* (1979), by Hernán Rodríguez Castedo—of which only a handful appear in the official dictionary. The topic of Spanglish was not at the core of his speech, but it was fully explored by many others present at the event. The purist faction talked about the dangerous threat that Cervantes's tongue faces today. But the more liberal voices, like that of Mutis, the author of *The Adventures of Maqroll*, took a stand. Mutis said that to reject Spanglish denounces a sort of "inexcusable innocence" and that the language should not fall prey of "la conspiración de los zombies."

To fully understand Spanglish, the history of English and its acceptance by the academy needs to be taken into consideration too. The exodus of the barbarous Saxons and Jutes into Britain around 450 A.D., and their interaction with the Celts and the Normans gave room to the language known as el inglés, but the process was unhurried. By the time *The Canterbury Tales* by Geoffrey Chaucer came along, it had undergone a normalized syntactical course, but some grammatical patterns were still untouched. *Cawdrey's Table Alphabeticall*, published in 1604, was the first attempt to offer a systematic approach to vocabulary. By the time Doctor Johnson embarked on his dictionary in the 18th century, there was still, in his own view, much to be cleaned: Gallicisms had a "pervasive" influence and authors, from Milton to Dryden, often spelled the same word in different ways. The English used today in Australia is not the same one employed in Nigeria, New Zealand, and India. But the differences are not only geographical: Nigerian English, as any other, is in constant evolution.

Let me now invoke an essay by Matthew Arnold called "The Literary Influence of Academies." Published in 1864 in *Cornhill Magazine,* in it Arnold praised Cardinal Richelieu and his fellow "enlightened" Frenchmen for forming a centralized governmental institution whose duty is to safeguard the French language—a "literary tribunal," Arnold called it, not only devoted to purifying and embellishing the vocabulary but, also, to serve as a body where "the works of its members were to be brought before it previous to publication, were to be criticized by it, and finally, if it saw it fit, to be published with the declared approbation."

Arnold believed this to be a positive approach, for it created "a form of intellectual culture which shall impose itself on all around." He was convinced England should follow in the same footsteps. In fact, Arnold's essay was una invectiva against the British character. "We all of us like to go our own way," he announces, "and not to be forced out of the atmosphere of commonplace habitual to all of us;—'was uns alle bändigt,' says Goethe, 'das Gemeine.' " Arnold wants standards in the lexicon and literature of his people. But he fails to realize that precisely the absence of a government-sponsored agency allows a free-flow of talent that, on its own rule and cadence, regulates quality in language and art.

He was mistaken . . . The fact that the English language doesn't have a soul-protecting body is reason to rejoice, as far as I'm concerned. For one thing, it is far more accepting of slang, and slang, as S.I. Hayakawa said in 1941, "is the poetry of everyday life."

La poesía de todos los días.

In a strict sense, there has never been anything similar to the *Diccionario de la Real Academia Española de la Lengua.* Attempts to make a dictionary have always been the result of individuals unaffiliated in political terms, such as Cawdrey, Blount, Kersey, Bailey, and Webster. Yet Doctor Johnson remains, undoubtedly, the magisterial model *por excelencia.* In 1746, at the age of thirty-six (and shortly after the last volume of *Autoridades* was made available in Madrid), he embarked on his *Dictionary of the English Language.*

In many ways, Johnson's project follows the same pattern in lexicography: to define a word is also to display examples, e.g., quotations of canonical figures whereby its usage is seen in the proper context. The story of the development of Johnson's project has been chronicled time and again by biographers, from James Boswell to Walter Jackson Bate. It is a story of patience and perseverance and of an individual's encyclopedic knowledge that triumphs over physical exhaustion and human inefficiency. Johnson proves Arnold wrong: his quest shows the individualism at the heart of Anglo-Saxon civilization—one man devoted for almost a decade, until 1755, when the book appeared, to codify la lengua de Shakespeare. His introduction begins poetically:

> It is the fate of those who toil at the lower employments of life to be rather driven by the fear of evil than attracted by the prospect of good; to be exposed to censure, without hope of praise; to be disgraced by, miscarried, or punish for neglect, where success would have been without applause, and diligence without reward. Among these unhappy mortals is the writer of dictionaries.

Johnson recognized that language is in constant mutation. Still, his mission, in his own terms, is to honor his country "so that we may no longer yield the palm of philology without a contest to the nation of the continent." He hopes "to give longevity to that which its own nature forbids to be immortal." He argues that elsewhere academies have been established to the cultivation of style, but is wary of one such entity in Britain. If it is ever established, he argues, he wishes not "to see dependence multiplied" and hopes that "the spirit of English liberty," el espíritu de la libertad inglesa, won't be hindered or destroyed. He believes the worst malady to inflict on a language is spread by translators too prone to use foreign words, especially French, rather than the colloquial alternatives. But his dictionary is not a bastion against foreign intervention. Al contrario, it establishes Greek, Roman and other etymologies by tracing the antiquity of a word, and it allows for neologisms that have in his eyes a *raison d'être*.

Even the buildup of the *Oxford English Dictionary*, by far the most reputed lexicon in the English-language orbit, commonly known through its acronym *OED*, serves as a paradigm of individualism. In 1857 Richard Chenevix Trench, then Dean of Westminster, called for the effort to put out a new dictionary to cure "the deficiencies of the language." Work began in 1878, and the actual publication of "125 constituent fascicles," orchestrated by fully committed scholars with the support of hundreds of people around the world, took place from 1884 to 1928. While the endeavor was dedicated to Queen Victoria and early copies were presented to King George V and to the President of the United States, it was, in all accounts, a non-official effort covered by Oxford University and Clarendon Press. That nonaffiliated status, and the objective to categorize words from English-language regions far and wide, remain intact. Like the *Diccionario de la Real Academia*, as time goes by the *OED* includes terms deemed improper or obscene at first.

In short, the difference between lexicological acceptance and rejection of the Spanish and English languages in their respective milieus, entre la flexibilidad y el dogmatismo, is rooted in history. With the fever for independence in the Americas in the mid 19th century, and even before the first branches of the Spanish academy were established in the Americas—after Colombia came Ecuador, then Mexico, and so on, until 1955, when an academy was created in Puerto Rico, followed in 1973 by its equivalent in the United States: *Academia Norteamericana*—each of these branches have released their own lexicons of local Spanish, e.g., diccionarios de mexicanismos, peruanismos, colombianismos, argentinismos, and so on. They also contribute regularly to the matrix in Madrid by sending freshly coined terms to be considered for inclusion in the *Diccionario de la Real Academia Española de la Lengua*. Their individual contributions testify to the varieties of Cervantes's tongue on this side of el Océano Atlántico. Nevertheless, the edition prepared in Madrid remains the "official" and authoritative one and is sold regularly all across the Hispanic world. (The latest update was made in 2001.) In contrast, numerous dictionaries of English are published from Canada to Britain, from the Caribbean to Australia. They are not a federal tool. Their function is not to legislate but to record. And their relationship with the *OED* is diffuse at best.

ℭℛℴ

*A*lmost on a daily basis, we agonize over the death of another one of the thousands of languages in some region of the globe. The emergence of a new tongue, on the other hand, particularly one that is the result of *mestizaje* and also of American imperialism, is dressed in a shroud of controversy.

Por qué?

La gente se queja, una y otra vez.

Think of it this way: every minute another galaxy is born. Do those births threaten our own existence? Not in the least. What they offer is an opportunity to reflect on the origins and explanation of our own universe.

O no?

Dónde comienza el espanglés—where does Spanglish begin? Is it possible to identify its birth date with any precision?

Like that of any form of communication, its origins respond to the needs of the population that uses it. Between 1492 and the mid 19th century, the encounter of the two weltanschauungs, Anglo-Saxon and Hispanic, produced a bare minimum of verbal miscegenation. The chronicles of conquest and conversion of Alvar Núñez Cabeza de Vaca (about Florida, in particular), "El Inca" Garcilaso de la Vega, and Gaspar Pérez de Villagrá, Fathers Eusebio Kino and Junípero Serra, and many others, all have as their target someone living in the Iberian Peninsula. They are composed in a Castilian Spanish colored by few regionalisms. The picture changes dramatically in the Mexican Southwest between 1810 and 1848. Early on Napoleon sold Louisiana to the United States, and shortly after, the arrival of Anglo-Americans to Arizona, Texas, New Mexico, and California—slow at first so as not to anger the local and federal authorities—transformed the region.

Yes, la modernidad arrived!

Missions were secularized, the Santa Fe Trail was opened by William Bicknell in 1822 and trade became attractive in spite of exorbitant taxes by Mexican officials. It was in those years that Texas became Americanized, with a population that quadrupled between 1820 and 1830, mostly as a result of new Anglo arrivals. The dialogue of Spanish and English increased as an obvious consequence. By 1848, when the Treaty of Guadalupe Hidalgo was signed by the Mexican dictator Antonio López de Santa Anna thereby selling for $15 million—¡qué oferta!— two thirds of Mexican territory to the White House, the juxtaposition of cultures was extensive. With the treaty, the population that lived in those territories in Arizona, California, New Mexico, and so on, switched citizenship from one day to the next. Article VIII of the document in English is clear about the physical status of these people. "Mexicans now established in territories previously belonging to Mexico," it stated, "and which remain for the future within the limit of the United States, as defined by the present treaty,"

> shall be free to continue where they now reside, or to remove at any time to the Mexican Republic, retaining the property which they possess in the said territories, or disposing thereof, and removing the proceeds wherever they please;

without their being subjected, on this account, to any contribution, tax or charge whatsoever.

Those who shall prefer to remain in the said territories, may either retain the title and rights of Mexican citizens, or acquire those of citizens of the United States. But they shall be under the obligation to make their election within one year from the date of the exchange of ratifications of this treaty: and those who shall remain in the said territories, after the expiration of that year, without having declared their intention to retain the character of Mexicans, shall be considered to have elected to become citizens of the United States.

In the said territories, property of every kind, now belonging to Mexicans, not established there, shall be inviolably respected. The present owners, the heirs of these, and all Mexicans who may hereafter acquire said property by contract, shall enjoy with respect to it, guaranties equally ample as if the same belonged to citizens of the United States.

Curiosamente, no mention is made anywhere in the document of the inhabitants' "madre lengua," although it was said in newspaper reports that, since language is an inalienable civil right, "it shall be respected thoroughly." History is written not by the conquered but by the victors, though; before long, English became the dominant tongue of business and diplomacy in the newly acquired Southwestern lands. But the usage of Spanish in schools and the household did not altogether vanish. Newspapers such as *El Clamor Público* in Albuquerque and *El Nuevo Mundo* in San Francisco serve as testimony to the relevance of the tongue. *El hijo de la tempestad* by Eusebio Chacón, the oral history of Eulalia Pérez and the call-and-response *El trovo del viejo Vilmas y Gracia*, the anonymous *Los Comanches*, are examples of its vitality.

By the time the Spanish-American War unfolded, Key West and New York had become magnets of immigration and solid Puerto Rican and Cuban communities had their roots in them. But it was clear that, as the so-called American Century was about to begin, the communication code had changed. From 1901 until the end of the millennium, dictionaries of Anglicisms were published all across the Hispanic world—in particular in Mexico, Cuba, Argentina, and Spain, by scholars like Ricardo Alfaro, Washington Llorens, Elena Mellado de Hunter, and Juan José Alzugaray Aguirre—with more and more frequency. This, surely, is a symptom of the verbal cross-fertilization experienced from north to south. Words like *buckaroo, rodeo, amigo, mañana,* and *tortilla* made it into English; likewise, *gringo, mister,* and *money* begin to be used in Spanish. The Nicaraguan poet Rubén Darío, the anointed leader of the Modernista movement that swept the Americas between 1885 and 1915 and included luminaries like José Martí, Delmira Agustini, and Leopoldo Lugones, denounced in one of his poems the oppressiveness of the American language:

> *¿Seremos entregados a los bárbaros fieros?*
> *¿Tantos millones de hombres hablaremos inglés?*
> *¿Ya no hay nobles hidalgos ni bravos caballeros?*
> *¿Callaremos ahora para llorar después?*

My unpoetic translation:

> Will we surrender to the ferocious barbarians?
> That many millions of people will end up speaking in English?
> Are there no longer noble *hidalgos* or brave knights?
> Will we fall silent today in order to cry tomorrow?

*O*nce the historical context of Spanglish becomes clear, a number of absorbing *puntos* of comparison emerge. Ebonics, *por ejemplo.* Do the two have anything in common? Expressions like "I own know what dem

white folk talkin bout" and "Hey, dog, whass hapnin?" are not uncommon among Black youth, especially in urban centers across the country.
This pattern of communication follows has its own grammar and syntax.
It is, for the most part, a spoken language nurtured by oral tradition,
even though poets and novelists like Zora Neale Hurston, Alain Locke,
Joan Teomer, and Countée Cullen of the Harlem Renaissance in the
twenties, and scores of descendientes, among them Richard Wright, in
Native Son and La Toni Morrison with her *Beloved*, a book instrumental
in her receiving the Nobel Prize for Literature, have done wonders
through artistic transcription.

Also known as African-American English, and Black lingo, the
idiom, in the words of Geneva Smitherman, author of the book *Black
Talk* (1994), is "a complicated system. Made even more complex by the
existence of Euro-American patterns of English within the Africanized
English system." Indeed, there is little doubt that Ebonics is an intra-
ethnic slang used by members of a minority group to establish a bridge
of identity. Its foundation dates back to the age of slavery. And class, to
a large extent, marks it, as lower-class people especially in urban centers
embrace it. "As far as historians, linguists, and other scholars go," Smith-
ernman says,

> during the first half of [the 20th] century it was widely be-
> lieved that enslavement had wiped out all traces of African
> languages and cultures, and that Black "differences" resulted
> from imperfect and inadequate imitations of European Amer-
> ican language and culture. George Philip Krapp, writing in
> the 1920s, is one linguist who held this view about the
> speech of Africans in America. In the 1960s these opinions
> came under close scrutiny and were soundly challenged by a
> number of experts, such as the historian John Blassingame
> and the linguist J. L. Dillard. Today scholars generally agree
> that the African heritage was not totally wiped out, and that

both African American Language and African American Culture have roots in African patterns.

Spanglish también is often an intra-ethnic vehicle of communication,
though only en los Unaited Esteits. It is used by Hispanics to establish
a form of empathy between one another. But the differences with Ebonics are sharp: Spanglish, for one thing, is a result of the evident clash between two full-fledged, perfectly discernible lenguas; and it is not defined
by class, as people in all social strata, from migrant workers to upper-
class statements like congressmen, TV anchors, comedians, use it regularly. South of the Rio Grande, Spanglish also knows no boundaries as it
permeates all levels of the economic ladder.

The interface of Ebonics and Spanglish is especially strong en la
música. For music is the most fluid expression of the Latin soul, el alma
latina.

Not too long ago, I found myself inside a taxi amid una combustión
de tráfico in Manhattan. The direct radio contact between the driver
and his dispatcher was electrifying. The woman arranging pickups and
departures savvily switched from Spanish to English and back, injecting
her speech with florid Spanglishismos that belonged to neither tongue.
The driver was probably of Asian origin—in his features, evidently a non-
Latino. Still, he understood perfectly what the woman was telling him,
answering back in Spanglish. He had learned the jargon from her and
others in the company and was clearly comfortable with it. His accent
was evident. But this didn't impede the dialogue. Why should it? In fact,
as far as I was able to judge, nobody seemed to notice.

Then, at one punto, he tuned into a radio station playing Latin rap.
I recognized the song *"Locotes,"* by the popular group *Cypress Hill.*

Moving his lips, the driver attentively repeated several stanzas. Dis-
tinctively Chicano terms, such as *ese, clica, gabacho,* and *jale,* weren't alien
to his Chinese appetite. On the contrary, he relished them. Was the experience surreal? No, ciertamente no. Latin music—not only salsa but

rock, hip-hop and jazz—has spilled beyond its ethnic enclave. On their way to mainstream culture, many of those rhythms underwent an infusion of African-American talent. Ebonics and Spanglish are juxtaposed in them.

The other useful point of comparison to understand el poder y alcance del espanglish is Yiddish. Upon my arrival in New York City in the mid eighties, I sensed that strange affinity between Spanglish and Yiddish. The explanation might have to do with my upbringing: Yiddish, another mishmash of languages, was part of my early education in the small Jewish enclave in Mexico in which I came of age. Spanish and Yiddish cohabited at home and in school, and to a lesser extent, so did Hebrew. The interplay of tongues wasn't at all strange for me, my siblings, and friends. On the contrary, it was common: people chose the language they preferred based on the context in which they found themselves: school, home, the synagogue, the outdoors. What was uncommon, though, was the mixture: it was always Yiddish or Spanish, never Yiddish and Spanish. Still, Yiddish is the result of the crisscross of Hebrew and German, to which Slavic terms were added, and then some more from Polish, Russian, Romanian, Latvian, French, English, etc.

Is Spanglish the Yiddish of today?

Un poco sí, un poco no . . . At first sight, the equation appears to be absurd. But is it? Benjamin Harshav, in *The Story of Yiddish* (1984), does a commanding job in chronicling the odyssey of Yiddish from rejection to full embrace. The dialect—it has been rightly described as an "internal tongue," e.g., a tongue spoken by an ethnic group to distinguish itself from the environment—was used by Eastern European Jews for seven centuries, from the 13th until the 20th, when, in 1945, the Auschwitz chimneys systematically killed it. Its linguistic sources are plentiful. It was first known as a gibberish for women and children and was looked down by rabbis and the intelligentsia as unworthy of Talmudic dialogue. Nevertheless, by the 19th a vast majority of poor, uneducated Jews, male and female alike in the so-called Pale of Settlement that included Poland, Lithuania, and Galicia, no longer were fluent in Hebrew and only spoke Yiddish, which time had turned from a jargon to a dialect and then into a mature language. (By the way, in Yiddish the word *Yiddish* means "Jewish.") So around 1865 Sh. Y. Abramovitch, the grandfather of Yiddish literature, made the conscious decision to author his novels and pedagogical treatises in it.

In *"Mayn Lebn,"* an autobiographical essay anthologized in his *Complete Works*, Abramovitch wrote: "I tried to compose a story in simple Hebrew, grounded in the spirit and life of our people at the time. At that time, then, my thinking went along these lines: Observing how my people live, I want to write stories for them in our sacred tongue, yet they do not understand the language. They speak Yiddish. What good does the writer's work and thought serve him, if they are of no use to his people? For whom was I working? The question gave me no peace but placed me in a dilemma." From then on it took enough self-confidence to generate masters like Sholom Aleichem, Peretz, Ansky and even Marc Chagall, whose pictorial images are but translations of his *shtetl* background. Play, stories, novels, poems, commentary, and translations were done in it. It vanished eventually, but not entirely—nunca para siempre. In 1978, one of the Yiddish literati, Isaac Bashevis Singer, a native of Poland and a New Yorker by choice, was awarded the Nobel Prize for Literature.

There was never one Yiddish but many: *galitzianer, litvak,* etc.

Yes, Spanglish shows the characteristics of an internal tongue, at least in the United States: it is often used by Latinos to define their own turf. But it has many other uses too: it is a transitional stage of communication in the process of English-language acquisition, it is a fashion, too. But in Latin America and the Caribbean the category of lengua intra-étnica, internal tongue, falls apart altogether, and another set of categories is brought to the fore: margin vs. center, imperial culture vs. colonies, etc. Also to consider is the fact that the presence of Yiddish in America is a result of the Eastern European immigration in the second part of the nineteenth century and the first twenty years of the next. Its relevance was huge in the Lower East Side of New York, among other places, where newspapers like *Der Forverts* had a wide readership.

Leo Rosten, in his classic *The Joys of Yiddish* of 1968, makes fun of the chaos—yes, *mishmash*—between Yiddish and English. "It is a remarkable fact," Rosten argues, "that never in its history has Yiddish been so influential—among Gentiles. (Among Jews, alas, the tongue is running dry.) We are clearly witnessing a revolution in values when . . . the London *Economist* captions a fuss over mortgage rates: 'HOME LOAN HOOHA.' Or when the *Wall Street Journal* headlines a feature on student movements: 'REVOLUTION, SHMEVOLUTION.' Or when a wall in New York bears the eloquent legend, chalked there, I suppose, by some derisive student in English: 'MARCEL PROUST IS A YENTA.' " "Get lost!" "O.K. by me," "I need it like a hole in the head!" and "Who *needs* it?" are all expressions that come from the Yiddish.

Spanglish too is looked down at by intellectuals north and south of the Rio Grande. Still, authors like Piri Thomas, Giannina Braschi, Juan Felipe Herrera, Sandra Cisneros, and Junot Díaz regularly employ it to explore the emotional depth of their characters.

"See, I tol' yuh," says un judío, a Jew, in a conversation between two nuyorkinos in Crown Heights.

"But I hav sumtin tel yuh," replies his Dominican neighbor, leading him to a building across the street. "Du yu no was the bes wor for a person that tacs two or mor lenguas?"

"No," answers el judío.

"Multiparlante . . ."

"So?" adds el judío.

"An was the bes wor for a person that tacs one lengua?" asks the Dominican.

"I doh no . . ."

"Un gringo . . ."

Naturally, the usage of slang by American authors isn't new: Mark Twain indulged in it masterfully in the 19ᵗʰ century, and scores have followed his example. This doesn't mean that what Huck Finn says is to be taken as a syntactical lesson. True, pero el hecho de que existan tales ejemplos, the fact that literature is malleable enough to reflect the street practice serves as a form of legitimization. When Yiddish literati repli-cated the jargon of the folk, they were hailed as "representatives of the collective soul."

In any case, Spanglish, to be fully understood, ought to be compared to Ebonics and Yiddish. Not that they are the same, but they have ele-ments in common. There are dramatic differences too: to suggest that Spanglish is the Ebonics of and Yiddish of Latinos es un error, of course.

Still, to understand the development of Ebonics and English, and for that matter also those of el español y el inglés, serves as window to ap-preciate Spanglish.

In the quest for completeness, it is also useful to think of varieties of Spanglish in strategic areas beyond the American continent and even hemisphere. In the Rock of Gibraltar, which for ages lived under Moor-ish domination but since the Treaty of Utrecht in the 18th century is in the hands of Britain "forever" (much to Spain's dismay), a mixture of English and Spanish is used. And in Zamboanga City and other south-western parts of the Philippines, Chavacano is preferred, a dialect with about 90% Spanish words that uses Tagalog-style grammar and syntax— a sort of Spangalog.

Other galaxies, our own.

◦◦◦

*M*y incessant curiosity for Spanglish has brought me scores of ene-mies. Since I first taught my course *The Sounds of Spanglish,* a sort of frenzy has taken shape. The BBC and NPR sent reporters into the classroom and then pursued the story in other vicissitudes. The Madrid newspaper *El País* broke the news that I had been compiling a Spanglish dictionary with an interview that caused dismay among purists. It then devoted another feature to the course itself, describing it as "la primer cátedra de Spanglish en el mundo." Opinion pieces started to appear in short order in *ABC* and *La Vanguardia* in Spain, and in countles dailies and weeklies in the Americas. In the English-speaking world, *The New York Times* and *The (London) Times* spread the news, and the echo was heard as far away as Italy, Brazil, Germany, and France. An Internet attacker described me as the Cheech and Chong Professor of Spanglish.

In the United States, a deluge of angry syndicated pieces and obsecene e-mails started to appear—and it only seems to increase in quantity. Hundreds have arrived in all shapes and tones. A typical one, by Rodrigo Díaz del Vivar (I've changed the name, again), reads:

> Me da asco saber que hay personas como usted que se siguen empeñando en tratar de acabar con un idioma tan hermoso como lo es el español. No puedo creer que haya en el mundo personas que quieran seguir protegiendo el supuesto espanglish que buen daño le hace al idioma.
>
> ¿A usted le gustaría que sus hijos hablaran espanglish en lugar de hablar un correcto español? Creo que no, pero ya me quedan mis dudas de qué tan brillante puede ser una persona que defienda semejante atentando al idioma.
>
> No sé de dónde salió semejante monstruo, pero lo que sí sé y lo tengo seguro es que es un anti-hispano como lo son muchos americanos, y no es porque odie a los hispanos, sino que está atentando contra nuestro idioma.
>
> Qué desgracia tener personas como usted dentro de la comunidad hispana.

My own English translation:

> I'm repulsed to know that there are people like you still devoted to the destruction of as beautiful a language as Span-

ish. I can't believe there are people in the world who insist on protecting the supposed Spanglish that in some measure injures our tongue.

Would you like your children to speak Spanglish instead of a correct Spanish? I don't think so, although I've doubts of how brilliant a person who defends such an attempt against the language can be.

I don't know where such a monster like you came from, but what I do know and I'm sure of is that you're an anti-Hispanic, the way many Americans are, and it isn't because you hate Hispanics, but because you're attacking our tongue.

What a pity to have people like you in our community.

The majority of these attacks—approximately 95%, in my estimation—are in Spanish. This, I believe, is symptomatic. In the Iberian Peninsula, the spread of el espanglés has become a national obsession. An army of commentators believes that its vitality is an omen: Hispanic civilization on this side of the Atlantic will survive in the future, but in a drastically altered, almost unrecognizable form. It is left to the Spanish government today, as well as to educators, to push for a serious study of el español en los Unaited Esteits. The obscenities have also come from official agencies. When news of my compilation of Spanglish terms spread throughout the Hispanic world, the *Academia Norteamericana de la Lengua,* which remains but a branch of its Madrid headquarters, issued an open letter—a fatwah, as an interneta described it once, portraying me as "el Salman Rushdie de los latinos"—denouncing the effort as an affront.

In the Americas, this reaction is far less palpable. Perhaps because people are accustomed to being colonized by foreign powers—Spain included—Spanglish is perceived as an attractive mixture that announces the emergence of a new self hispano.

Among English speakers north of the Rio Grande, the debate has less to do with imperialism than with assimilation. Spanglish, the purists suggests, is the result of a bankrupt system of Educación Bilingüe: when teachers and parents forget how to delineate the line between one language and the other, the outcome is verbal chaos. Other reasons are added to its existence, among them a general state of "laziness" among Hispanic immigrants to learn proper inglés completamente and the endorsement of multicultural programs that encourge cultural hybridity rather than discourage it.

Among conservative Latinos—and there are numerous—Spanglish, as the e-mail I quoted above proves, is a form of shame. Other inmigrantes assimilated into the Melting Pot by sacrificing their original tongue, and so should we.

But should we?

Although I'm perfectly aware of my public role as the target in the animosity against el espanglés, I've chosen silence as a response to the criticism. The reason is uncomplicated. The attacks are a manifestation of a buried emotional reaction. Unfortunately, Hispanic civilization has never quite understood the role of constructive analysis. To study Spanglish isn't to endorse its future, thus undermining el español . . . On the contrary, to scrutinize is to better understand where we come from and who we are.

As I've stated above, in no way do I disagree with those who believe that Spanish and English should be spoken well—hablémoslos como se debe. My interest in Spanglish isn't a question of either/or: Professor Stavans, Do you prefer Spanglish instead of . . . ? No, there aren't any *insteads.* . . . Me gustan todas por igual, I like them all the same.

I realize, obviously, that for many impoverished Latinos the possibility of speaking el inglés, el español and Spanglish isn't really an option. But should that curtail our constructive analysis? And for how long will the so-called educated insist that los pobres don't speak a tongue, they simply destroy it? Class is an integral component of our way of life. When it isn't Spanglish we're complaining about as we address those that we portray as less worthy than us, it is something else: as Ambrose Bierce puts it, slang is invariably a disgrace—"the grunt of the human hog *(Pignoramus intolerabilis)* with an audible memory."

(By the way, my children do speak espanglés!)

⟨⟨ҩ⟩⟩

I tol yuh . . .

Not long ago, as I walked the streets of San Antonio, I came across a sign that read: "Se prohibe hanguear!" Weeks later, in La Loisaida, as the Lower East Side of Manhattan is called by Nuyorricans, I read the same sign, although spelled in a different way: "¡No jangear!"

I was thinking about this coincidence, about the unconscious effort to standardize Spanglish, when I stumbled upon Lisa Martínez. I was descending the staircase of a subway station while she was coming out from it . . .

After a few seconds, I heard a recognizable sound: Wasá?

The two of us were overwhelmed by a sense of disbelief. How long had it been since we last had seen each other?

A smile, a brief cómo te va. Then I asked her if she had time to spare. Happily, ninguno de los dos was in a rush. So I invited her to have a coffee.

Lisa briefed me about her odyssey after she left New England. She said she first moved back to Istlos, where for months she devoted her-self to community service. Over time, she did complete a BA degree in a local institution, then enrolled en la escuela de leyes, Law School, this time en Nuyol.

Why come back to the East Coast? I asked, aware of her loyalty to California.

"Power is on this side, profe . . . ," she replied. "You know it, right?"

We spent approximately half an hour together. Only in the last few minutes did the topic become Spanglish.

"Y usté, I know you sigue con el asunto del espanglés! I've heard you on the radio. And I saw you interviewed on Spanish TV . . . It was culísimo!"

I told her I had translated the first chapter of *Don Quixote* and that I was about to publish a Spanglish dictionary.

"You opened the door for me, Lisa. Before that fateful day in which you came to la oficina to say adiós, I had not taken Spanglish to heart."

"Yeah, I smiled when you began talking in la jerga loca to me . . . It sounded so foreign, so artificial. It didn't seem yours at all!"

There was a pause.

"By the way, profe. How would you translate the sentence 'Entre, entre y tome asiento' into English?"

I laughed. She was referring to a line from the movie *El Super* (1983).

"Between, between and drink a chair," I said.

"Hey, you're getting good!"

Lisa Martínez then sang for me a couple of stanzas of the so-called "Official Spanglish nacional Anthem," which she had found on the Internet. "It's about the Boricuas, profe. . . ," she announced. "Like the ones together with whom we made the alboroto about Spanglish in your class. Se acuerda?"

I heard her versify:

> *Y paaaaa'*
> *El carajo with the numbers*
> *If I can't fly I'll swim*
> *Straight from El Barrio*
> *Back to Puerto Rico*
> *(Island by the sun blessed*
> *Island I never left*
> *I will settle there next)*

Así es how it must be
For the whole family
Dice our destineeeee!
De weather wasn't nice
Comfort cost a high price
Unlike in Puerto Rico
We kept cooking the rice
And re-heating the beans
And making cuchifrito.

I had a cappuccino y una bagel with cheese. Lisa drank jugo de china. We walked back to the subway and, in normal English, said good-bye. As I took the train to the Upper West Side, where I was planning to stay the night at a friend's house, I pondered how much I had changed in the last decade or so, and how much my surrounding had undergone a transformation too. "Pollito Chicken" by Ana Lydia Vega had been the first short-story I had encountered in the jargon. I had come across scores of others in recent years, not only by Latinos in los Unaited Esteits but also south of the border. In fact, just a few days ago I had seen in the magazine *El Andar* a powerful personal memoir by an Argentine, Susana Chávez Silverman. "Crossing the Riachuelo in a smudge-windowed bus, over into Avellaneda. Provincia de Buenos Aires. Roof patios, si se les puede llamar así (porque de encanto y relax tropical no tienen nada) . . ."

Also, a former student of mine, on a trip to El Viejo San Juan, had found a restaurant, *The Parrot Club* (address: 363 Fortaleza; telephone: [787] 725-7370), where the menu is entirely in Spanglish. "You should stop by on your next viaje, Ilan. Try the Jumbo Spiced Camarones, pan fried y servidos con bacon, tomato salsa, arroz blanco y jícama salad . . ."

And in the magazine *Latina* I had seen an advertisement for the US military service. It sought to attract Latinos: "Yo Soy el Army," it read in large-size letters.

Indeed, at times it seemed to me as if the world entire, el mundo entero, existed en Spanglish. The cacophony of voices that had attracted my ear in Manhattan in the mid eighties, at my arrival as an immigrant, had expanded its base of operation. Department stores, agencias de viajes, restaurantes, bucherías, auto repairs . . . el espanglés was used often and everywhere.

Might a monolingual feel disoriented while walking the streets? If so, does anyone care? Not in the least, it seems.

And does anybody pay attention to the rules of syntax? But again: does it matter?

This delicious—and delirious—mishmash is what Latino identity is about: the verbal mestizaje that results from a transient people, un pueblo en movimiento. One day, not too far into el futuro, a masterpiece in literature might be composed in that mishmash, one that will forever change the way we comprehend our world. It might take time for readers to appreciate it in full, but the existence of such readership is not improbable. I once heard Ana Celia Zentella, a specialist in Nuyorrican grammar and the author of *Growing Up Bilingual* (1997), describe the individual Spanglish-speaker as "two monolinguals stuck at the neck." It is a haunting, beautiful image that makes me think of Robert Louis Stevenson's *The Strange Case of Dr. Jekyll and Mr. Hyde*—one body, two selves.

Why do Spanglish speakers say rufo and not techo? Maybe because, as the Mexican intellectual Alfonso Reyes once said, "the law of the easiest effort" ruled the formation of language, and language is the freest of human acts. "Grammar is neither a useful art," Reyes added, "nor does it really teach us how to speak. . . . It is simply the rhythm we use as we formulate thought."

Yes, rhythm and thought—el ritmo y la razón.

A User's Manual

Dictionaries are word-books. The term *dictionary* itself comes from Medieval Latin: *dictionarium: dictio,* a word + *arium,* the suffix that implies compendium. Language, in constant movement, is impossible to capture. Any dictionary that prides itself as accurate must, to begin, recognize that the lingo on the streets will never be imprisoned. Spanglish, thus, creates a daunting series of tasks for the lexicographer. It is, to this day, an oral vehicle of communication. Many of its words and phrases have never been written down before. This means that the spelling is evasive, and frequently the etymology is too.

Do lexicons tell people how to speak? Or is it the other way around? There's a difference between the *normative* and the *descriptive:* the former regulates and systematizes, whereas the latter simply records. This volume is evidently descriptive. It isn't a didactic tool—its purpose is not to teach people *how* to speak Spanglish, but to represent the multifarious tongue(s) spoken by Latinos, and to a lesser extent by Hispanics the world over, at the dawn of the third millennium. The research

that has gone into this volume involved endless hours of reading periodicals and literature from the 19th century to the present. It also includes
research into the varieties of Spanish spoken in the United States since
the Treaty of Guadalupe Hidalgo was ratified in 1848. And it benefited
from the support of endless correspondents across the continent, as well
as in Argentina, Peru, Venezuela, Colombia, Guatemala, Mexico, the
Caribbean, and Spain. Last but not least, I scrutinized dictionaries published from 1876 to the present throughout el mundo hispánico. Scores
of voices recorded here have appeared before in reference manuals, at
times with a different spelling. I made ample use of these manuals, all of
which are listed in the bibliography of sources and material for further
reading at the end of the volume. I hereby express my wholehearted
gratitude to the editors.

Pero vayamos a lo nuestro: an explanation is in store to the user of
this lexicon. What does it contain? How is the material organized?
What methods were used to record the Spanglish voices? Let me begin
with a comment on the tension between the norm and its deviations in
the linguistic process. The spelling I have in every entry is the one
most commonly used in popular culture. In the quest for standardiza-
tion, I have consulted with dozens of linguists, lexicographers, gram-
marians, educators, and critics. *Spanglish: The Making of a New American
Language* includes thousands of entries that emerge from different national groups: Mexicans, Cubans, Puerto Ricans, Dominicans . . . all in
the mainland United States. The majority of these entries have crossed
boundaries, are already pan-Latino, and thus are no longer geographically specific. But geographical differences also prevail. To emphasize
them, I've accompanied some entries used in specific contexts with the
following designations:

Chicano {Ch}
Cuban-American {CA}

Dominican-American {D}
East Los Angeles {ELA}
Los Angeles {LA}
Massachusetts {Mas}
New Mexico {NM}
New York {NY}
Puerto Rico {PR}
Southwest {SW}
Texas {T}

A small number of voices are from countries other than the United States. In those cases, I use the following abbreviations:

Argentina {Arg}
Central America {Cen}
Chile {Chi}
Colombia {Co}
Cuba {C}
Mexico {Mex}
Spain {S}

Approximately 10 to 15% as of the entries in this volume belong to the category of Cyber-Spanglish. The abbreviation {CS} follows those voices that belong to this category.

This, needless to say, is a diccionario vivo. Spanglish is alive, mercurial, in constant mutation. It remains an oral code of communication. In the years involved in the preparation of the material in display here, an armada of students and I recorded and transliterated numerous words. Their pronunciation varied from one generation to another, as well as among communities and regions. The transcription has been a particularly difficult task. To merit inclusion here, a voice needed to be recorded—orally or in written form—at least three times. The research team also came across countless written words, but never as many as those only found in spoken form. I have done everything possible to offer alternative spellings that reflect the origin of the words alphabetized in this volume. When more than one is available, the alternatives are listed independently of each other, as in these cases:

bróquer (BRO-ker), n., m., 1. financial and real-estate advisor. 2. intermediary E»S. Also BROKER.

and

broker (BRO-ker), n., m., 1. financial and real-estate advisor. 2. intermediary E»S. Also BROQUER.

What counts as a Spanish word? [1] loanwords, such as *fiesta* and *arroyo* in English, and *advertising* in Spanish. [2] mistranslated terms, such as *banco* in Spanish for river bank, and limón in English. But more significantly, [3] adapted words like *estylo, jazear,* and *washatería,* which are often the result of misspellings and misunderstandings. It is not secret that false synonyms and antonyms, uneducated guesses, and egregious errors are the source of the people's vocabulary. The majority of the entries clearly fall in the category of "barbarisms" used by lexicographers: expressions pushed into an extreme—deformed, perverted by the speaker: *liquiar, marqueta, taipo* . . . Is there anything wrong with that? Not at all. After all, the language is the most democratic of human endeavors: it is *by* the people and *for* the people. A borrowing and an error repeated today become linguistic rules tomorrow.

An attentive reader will notice that Spanglish, at least as it is represented here, is not a symmetrical two-way street. Its speakers tend to adapt English words (mostly verbs and nouns) into the syntactical pattern of Spanish. This is as it should be, for Latinos in los Unaited Esteits are the ones at the linguistic avant-garde. They are in charge of revolutionizing language through sheer improvisation.

In the pages that follow, a typical entry reads:

> **laptop** (LAP-top), n., f., portable computer. "Me llevé la laptop to the *trip*." {CS} E»S Also LAPTOPA.

The word **in bold** is the Spanglish term as it has been recorded. In parentheses is its pronunciation in English. Next comes its grammatical nature. In this category, I've used the following abbreviations:

adj. Adjective
adv. Adverb
n. Noun
v. Verb
exp. Expression
interj. Interjection
prep. Preposition

When either a noun or an adjective is entered, the qualifiers "m." for masculine and "f." for feminine follow. These denote the usage of that noun in the Spanish-language context. The designation m/f signifies that the noun is used transgenerically. If the voice ends with a vowel "a" or "o," it is assumed that it needs the other vowel to be used in the opposite gender. Take this example:

> **aficionado** (a-fee-zio-NA-do), n., m/f., enthusiast. The term dates from 1845. S»E

Even though it is common in English to use the masculine *aficionado*, in Spanglish it is possible to use the voice as either *aficionado* or *aficionada*.

As a counterpoint to the Spanglish term, entries occasionally include the appropriate English and Spanish root. Or else, the original meaning of a word is given at the end of a sentence, as in these instances. For example:

> **adió** (a-DIO), n., m., goodbye. Sp. *adiós*. S»E

and

> **rewin** (re-WEEN), n., m., mechanism by which a tape is rolled backwards. Eng. *rewind*. E»S

Voices from the English used in Spain and the Americas are followed by the sign E»S. Conversely, voices originally from the Spanish-speaking world that have been adopted in the United States have an S»E. Terms changing meaning in the transaction have the various meanings listed.

I've also made an effort to include model sentences to illustrate the meaning of specific voices whose meaning is only understood in the proper context. For instance:

> **biciar** (bee-SEE-ar), n., m., V.C.R. "Ahora que Chucho me enseñó cómo usar el biciar, I never miss a soap opera episode." E»S

So as to not turn this volume into an illustrated guidebook, I've constrained myself from over-indulging in examples. The ones I've inserted are delivered in Spanglish and juxtaposing inglés y español.

When available, the date in which a word was first recorded is offered. Variant definitions are organized sequentially by order of the frequency in their usage.

Often appearing in the entries is a year date. This announces the first time the term was recorded in English. The principal, although not the sole, source of information is John Algeo: "Spanish Loanwords in English by 1900," included in *Spanish Loanwords in the English Language: A Tendency Towards Hegemony Reversal.* (A complete list of sources and material for further reading appears in the bibliography at the end of this volume.)

Also, Spanglish doesn't appear to use inverted question and exclamation marks. Thus, it is "Fuiste al bloaut sale de Macy's?" and not "¡Viva la barbarie!" And there is a plethora of voices that start with **Ch,** which might explain the absence of entries in the letter **X.**

Finally, a word about the overall ambition behind the pages that follow. Qué tan completo is this dictionary? By way of an answer, let me invoke the famous Spanish saying: ni están todas las que son, ni son todas las que están—the voices included here represent a fraction of the roughly 6,000 I've been able to trace over the years. I've mostly left out those whose occurence to the best of my knowledge remains sporadic. And only a small number of Spanglishismos from Central and South America have been incorporated.

My hope is to allow the lexicon to grow, que crezca y se reproduzca in future editions.

Lexicon

SPANGLISH » ENGLISH

A

@ (AT), prep., at. {CS}

abajar (ah-BA-xar), v., to descend. "Abaje your head." Sp. *bajar.* {SW}

abaldonar (a-bal-DO-nar), v., to abandon. Sp. *abandonar.* {SW, NR}

abismal (a-bis-MAL), adv., abysmal.

abocado (a-voo-KA-do), exp., assimilationist. Used as equivalent to Uncle Tom. Orig. a pear-shaped fruit. From Nahuatl *ahuacatl.* The term dates from 1697. Also AVOCADO.

abombado (a-bom-BAH-do), adj., m/f., overwhelmed.

abombamiento (a-bom-ba-MIEN-toh), n., m., state of overwhelmed activity. "Qué abombamiento!"

absentismo (ab-sen-TUIS-mo), n., m., absenteeism. Sp. *ausentismo.*

abstracto (abs-TRAK-toh), n., m., synopsis, abstract. "Dame el abstracto."

abusón (a-bu-ZON), adv., abusive. {Ch}

académico (a‑ka‑DE‑mi‑ko), n., m./f., 1. scholar. 2. adj., m/f. scholarly. "He is muy académico."

accento (a‑XEN‑to), n., m., accent. Sp. *acento*. {Ch} E»S

accesar (a‑XE‑zar), v., to access. {CS} E»S

ace (EIZ), n., m., rapid ball. Tennis term. "Viste el ace?" Also AZ.

aceitar (a‑ZEI‑tar), v., to accept. {Ch} E»S

acequia (a‑ZE‑kia), n., f., irrigation canal. The term dates from 1896.

ácid (AH‑cid), n., m., 1. amphetamine; 2. rhythmic description. "Me encanta el ácid jazz." Sp. *ácido*. E»S

acordando (a‑kor‑DAN‑do), adv., according, accordingly. E»S

adenda (a‑DEN‑dah), n., f., addendum. E»S

adió (a‑DIO), n., m., goodbye. Sp. *adiós*. S»E

adobe (a‑dob‑ee), n., m., mud and straw mixture used for architecture. The term dates from 1748. {SW, Sp}

adrés (a‑DRES), n., f., address. {CS} E»S

advertising (ad‑ver‑TAY‑sing), n., m., publicity effort. "Jaime trabaja en advertising." E»S

aeróbica (ay‑RO‑bi‑ka), n., m., dynamic female. E»S

aeróbicos (ay‑RO‑bi‑cos), n., m., aerobics. {S} E»S

afeccionado (a‑fe‑zio‑NA‑do), adv., overly emotional "Viste que es afeccionado? Did you see how emotional he is?" E»S

aficionado (a‑fee‑zio‑NA‑do), n., m/f., enthusiast. The term dates from 1845. S»E

aflictar (a‑FLEEK‑tar), v., to afflict. Sp. *afligir*. E»S

afluente (a‑flu‑EN‑te), n., m/f., well‑to‑do. Sp. *adinerado*. {Mex} E»S

aformencionar (a‑for‑men‑TIO‑nar), v., to aforemention. E»S

afterauers (af-ter-AU-ers), n., m., after hours. "Vas a la fiesta de after-auers?" (S) E»S

aftersheif (af-ter-SHEIF), n., m., aftershave. (S) E»S

aftersun (af-ter-SUN), n., m., mousturizing lotion. (S) E»S

agardente (a-gar-DEN-te), n., m., fiery spirits. Also AGUAR-DIENTE.

agente (a-GEN-te), n., m/f., secret agent. (CS)

aguakear (a-wa-KEAR), v., to awake. E»S Also AGUAKIAR, AGUA-QUEAR, AGUAQUIAR, AWAKEAR and AWAKIAR.

aguakiar (a-wa-KEAR), v., to awake. E»S Also AGUAKEAR, AGUA-QUEAR, AGUAQUIAR, AWAKEAR and AWAKIAR.

aguaplanin (a-gua-PLA-nin), n., m., high-speed water-skiing. (S) E»S

aguaquear (a-wa-KEAR), v., to awake. E»S Also AGUAKEAR, AGUAKIAR, AGUAQUIAR, AWAKEAR and AWA-KIAR.

aguaquiar (a-wa-KEAR), v., to awake. Also AGUAKEAR, AGUA-KIAR, AGUAQUEAR, AWAKEAR and AWAKIAR.

aguardiente (a-GUAR-dien-te), n., m., strong liquor, also known as Taos Lightnin. E»S Also AGARDENTE.

águate (A-gua-TE), interj., exp., be aware! "Aguate of the dog." (SW) E»S

agüitado (A-wee-TA-do), a., m/f., 1. sad, melancholic 2. to have inter-nalized bad news. "Está bien agüitado since his lady left him." (M, NR, SW)

airbag (AYHR-bag), n., f., airbag. E»S

airbus (AYHR-bus), n., m., airbus. (NR) E»S Also AIROBU and GUAGUA AEREA.

airlain (AYHR-lain), n., f., air company. E»S

aisber (AYHS-ber), n., m., 1. to treat someone coldly. 2. as an idiomatic expression: 'the cold shoulder.' "Le dio el aisber. He gave her the cold shoulder." {CA} E»S

ajax (a-JAX), n., m., clean, spotless. {CA} E»S

alambrista (a-lam-BREEZ-ta), n., m/f., illegal immigrant. {Ch}

albatea (al-ba-TE-a), n., m/f., well-read, literate person. "The college lo hizo albatea." {Ch}

albatros (al-BA-tros), n., m., a large sea bird. The term dates from 1681.

álbum (AL-boom), n., m., musical recording. E»S

alcaide (al-CAY-de), n., m., mayor or governor of a region. Sp. *alcalde.* {SW}

alcaidía (al-cay-DEE-a), n., f., region governed by an alcalde. Sp. *alcaldía.* {SW}

alegado (a-LEE-gah-do), n., m/f., alleged person. E»S

alegría (ah-le-GREE-ah), n., f., pigweed, bright red herb used as a cosmetic, chewed, and pressed for juice. In Sp. *alegría* means happiness. {SW}

alforge (al-FOR-geih), n., m., wide canvas or leather saddlebag. Sp. *alforja.* {SW}

alibi (a-LEE-bee), n., m., alibi. "He doesn't have ni un alibi." E»S

aliveo (a-lee-VE-oh), adv., lively, sprightly.

alrandom (al-RAN-dom), n., m., randomly. "Voy a optar entre las dos casas alrandom." {G} E»S

amá (AMAH), n., f., mother. {Ch}

amigoization (a-mee-go-IZA-tion) n., f., the process of Mexicanization of the U.S. Southwest. {SW}

amontillado (a-mon-tee-YA-do), n., m., a type of sherry. The term dates from 1825.

analfayuca (a·nal·fa·YU·ka), n., m/f., illiterate. {CA}

analización (a·na·lee·za·ZEEON), n., f., analysis.

ancorman (an·KORHR·man), n., m., TV news personality. {S} E»S

ándale (an·da·LEEH), c., get going. Sp. *andar.* Used with modifications in Mexico and Central America. {Cen, Ch, Mex, SW}

andaumen (an·DAU·ment), n., m., endowment. E»S

angélica (an·GE·lee·ka), n., f., unmarried girl. The term dates from 1885.

anion (a·NYON), n., f., onion. {NR} E»S

anotar (a·noh·TAR), v., 1. anotate. 2. comment. Sp. *apuntar.* {CS} E»S

antibaby (an·tee·BAY·bee), n., m., birth control pill. {S}

anticiclón (an·tee·cy·CLON), n., m., high pressure zone. {NR, PR}

antifrizer (an·tee·FREE·zer), n., m., fornication. {CA} E»S

apá (APAH), n., m., father. {Ch}

aparatichi (a·pa·REE·tee·tchee), n., m., a C.I.A. agent. {CA}

aparcadero (par·kea·DE·roh), n., m., parking garage. Also PAR·CADERO and PARQUEADERO.

aparcar (a·par·KAR), v., to park. E»S Also PARQUEAR, PAR·QUIAR, PARKEAR and PARKIAR.

apartamento (a·par·ta·MEN·toh), n., m., apartment. Sp. piso, departa·mento. "Vive en un apartamento amueblado en Manhattan."

aparthotel (a·part·HO·teel), n., m., hotel apartments. {SM} E»S Also APARTAOTEL.

apelación (a·peh·la·ZION), n., m., name. "Su apelación is Gonzalo." E»S

aplicación (a·plee·ka·ZION), n., f., application. E»S

aplicar (a·plee·KAR), v., apply. "Juana aplicó al college." {CS} E»S

archivador (ar-chee-va-DOOR), n., m., arhival. {CS} E»S

archivar (ar-chee-VAR), v., to store. {CS} E»S

armada (ar-MA-da), n., f., fleet of warships. The term dates from 1533.

armadillo (ar-ma-DEE-yoh), n., m., burrowing mammal with an armor of bony plates. S»E

armi (AR-mee), n., m., army. E»S

armitas (har-MEE-tas), n., f., type of leather chaps. {NR}

arrastrador (ah-rras-tra-DOR), n., m., machine gun.

arrastrar (ah-RRAS-trar), v., to crush.

arroyo (a-RRO-yo), n., m., 1. creek 2. creek bed. The term dates from 1597. {E}

arroz (ah-ROZ), n., m., rice. "The Sánchez family diet is arroz with frijoles." S»E

arrurrúz (a-RU-ruz), n., m., arrow-root, type of flour. {SW}

asetaloqueción (a-ser-tah-lo-kei-ZION), n., f., distribution of business material. {Chi}

asistén (A-SIS-ten), n., m., f., assistant. E»S

asistir (a-sees-TEAR), v., to help. Sp. *ayudar.* E»S

atachear (a-ta-CHEAR), v., to attach. {CS} E»S

atachmen (a-TACH-men), n., m., attachment. {CS} E»S

atexanado (a-te-xa-NA-do), adj., m/f., Texanized. {Ch} E»S

atlarg (at-LARG), exp., at large. "Hoy día permanecía atlarg el ladrón." E»S

atol (A-tol), n., m., corn porridge. Sp. *atole.* {M, SW}

atunar (a-TOON-ar), v., to harmonize. "Atuna tu guitarra! Tune your guitar!" E»S

auditear (au-Di-tear), v., to audit. {M, SW} E»S

auditor (au-DI-tor), n., m., auditor. {M, SW} E»S

auspicioso (aus-pee-ZIO-zo), n., m/f., auspicious. Sp. *audaz.* E»S

aut (AUT), exp., 1. out of fashion. 2. obsolete. 3. given a last chance. 4. Baseball term. "Esa pelota está out." E»S Also OUT.

autodate (au-to-DAIT), n., m., self-reminder. {CI} E»S

automotor (au-to-mo-TOR), n., m., car. E»S

automóvil (au-yto-MO-bil), n., m., car. E»S

avalancha (a-va-LAN-cha), n., f., avalanche. From Fr. *avalanche.*

averaje (a-ve-RAH-je), n., m., average. E»S

aviónica (a-VIO-ni-ka), n., f., flight science. "Es bueno para la aviónica. He knows how to fly." {PR}

avocado (a-voo-KA-do), exp., assimilationist. Used as the equivalent of an Uncle Tom. From the orig. a pear-shaped fruit, which comes from Nahuatl *ahuacatl.* The term dates from 1697. Also ABO-CADO.

awakear (a-wa-KEAR), v., to awake. E»S Also AGUAKEAR, AGUA-KIAR, AGUAQUEAR, AGUAQUIAR and AWAKIAR.

awakiar (a-wa-KEAR), v., to awake. E»S Also AGUAKEAR, AGUA-KIAR, AGUAQUEAR, AGUAQUIAR and AWAKEAR.

ay (AI), exp., alas. "Ay, she was right!"

az (AZ), n., m., rapid ball. Tennis term. "Viste el ace?" Also ACE.

Aztlán (az-TLAN), n., mythical Chicano homeland. {Ch, SW}

azucarcani (a-zoo-kar-KA-nee), n., m., cane sugar. {CA}

B

babay (bah‑BAIH), expt. Bye‑bye. E»S

babiche (ba‑BEE‑che), exp., son of a bitch. "Jaime es un babiche!" {Ch}
　　E»S

babún (ba‑BOON), n., m., African monkey. {C}

baby (BEI‑bee), n., m/f., baby. Sp. *bebé.* S»E

bacapear (ba‑kah‑PEAR), v., to back up. {CS} E»S

bacfid (BAK‑feed), n., m., backfeed. {CS} E»S

bach (BACH), n., m., batch. "Ellas trajeron un bach de documentos."
　　{CS} E»S

bachelor (ba‑CHE‑lor), n., m., unmarried man. E»S

bacín (ba‑SEEN), n., f., bassinet. "El enfermo necesita un bacín." {Ch}
　　E»S

backgámon (bak‑GAAH‑mon), n., m., backgammon. Board game. E»S

bácop (BAH-cup), n., m., saved computer information. "Seguro que tiene bácop of the program?" {CS} E»S Also BACUP.

bactrak (bac-TRAC), n., m., back track. {CS} E»S

bacup (ba-KUP), n., m., back up. {CS} E»S Also BACOP.

bagaje (bah-GA-cheh), n., m., baggage. E»S

baglady (bag-LAAY-dee), n., f., homeless woman. "Esa mujer que vive en las calles es una baglady." E»S

bagraun (ba-GROUN), n., m., background. {CS} E»S

baile (BAY-leh) n., m., party. "Es su birthday y dio un baile." Sp. *fiesta.* S»E

baipás (bai-PAZ), n., m., bypass. "In the surgical operation le dieron un triple baipás." E»S

baiquer (BAI-ker), n., m/f., biker. E»S

baisic (BEI-sik), adj., essential, basic. "Es baisic teoría." E»S

baisiquel (bay-se-KEL), n., f., bicycle. E»S

baiso (BAY-so), n., m/f., young person. {Ch}

bait (BAIT), n., f., bribe. From Mex. Sp. *mordida.* {Ch} E»S

baite (BAI-teh), n., m., computer term. {CS} E»S Also BAYTE.

baja (BAH-kha), n., f., steep descent. "El tuvo el accidente en la baja." {CA} S»E

bakupear (bah-koo-PEAHR), v., to backup. {CS} E»S

bakupeo (bah-koo-PEH-oh), n., m., the act of making a backup. {CS} E»S

balance (ba-LAN-ce), n., m., bank statement. E»S

balasto (BAH-las-tro), n., m., blast. "That is when ocurrió el balasto of 1939."

bambi (BAM-bee), exp., innocent female person. "No seas bambi with your new boyfriend." {CA} E»S

banana (bah-NA-nah), n., f., plaintain fruit. Sp. *plátano*. {CA}

bananaspli (BAH-na-nah-SPLEE), n., m., banana split, a type of ice-cream sundae. E»S

banco (BAN-ko), n., m., 1. river bank. In Sp. *banco* means bank. *Ladera* is the word for river bank. {Ch} 2. sperm donor. "He is un banco."

banda (BAN-dah), n., f. musical band.

bandido (ban-DEE-doh), n., m., outlaw. S»E Also BENDITO.

baner (BAH-ner), n., m., flag. "En la parada llevaban un baner de Puerto Rico." {CS} E»S

banners (BAH-ners), n., m., Web advertising. {CS} E»S

banquiar (ban-KEEAR), v., to bank. "At what times vas a banquiar?" {Ch}

baquiar (bah-KEEAR), v., 1. to go back. 2. to back up. "Me baqueo a casa after school." {Ch, CS}

barbakiu (bar-ba-KEEUH), n., f., BBQ. Sp. *barbacoa*. E»S

barcoud (BAR-cowd), n., m., bar code. {CS} E»S

barman (BAR-man), n., m., f., bartender. E»S

barracón (bah-rah-KOON), n., f., 1. warehouse. 2. slave quarters. Eng. *barracoon*.

barrio (BAH-ree-o), n., m., neighborhood. "Our barrio is nearby." The term dates from 1841. {Ch, SW} S»E Also HOOD and HUD.

barrio virtual (BA-ree-o VEER-tu-al), n., m., World Wide Web page devoted to Chicano issues. {Ch, CS, SW} Also VIRTUAL BARRIO.

bas (BAZ), n., m., bus. "El último bas left already." {CA, SW} E»S

basis point (bay-sees POINT), n., m., interest-rate points. E»S

basket (BAS-ket), n., m., basketball. E»S

basqueta (bas-KE-tah), n., f., garbage bin. E»S

básquetbol (bas-ket-BOL), n., m., basketball. E»S

bastardiar (bas-tard-DEEAR), v., to engage in extramarital sex; 2. to give birth to bastards. "Jaime is probably out bastardiando."

bastonero (bas-to-NEE-ro), n., m., master of ceremonies.

bat (BAT), n., m., baseball bat. E»S

bateador (bah,teha-DOOR), n., m., person at bat. Baseball term.

batear (bah-TEAR), v., to bat. Baseball term. {Ch} E»S

batería (bah-te-REE-ah), n., f., legal term of battery. "He got himself into a serious problem of batería." {CA} E»S

bato (BAH-toh), n., m., friend. {Ch} Also BRO, BRODER, BROTHER, CARNAL, GÜE, GÜEY, and VATO.

bato furris (BAH-toh PHOO-rees), exp., good-for northin'. {Ch}

baunsiar (BAUN-seear), v., to bounce a check. "A check of mine se baunsió."

bazuca (bah-ZOO-ka), n., f., bazooka. E»S

beat (BEET), n., m., musical rhythm. E»S

bedanbrec (BED-an-BREIK), n., m., bed and breakfast. E»S

beibisiter (bay-bee-SEE-ter), n., m/f., babysitter. E»S

beicon (BEY-kon), n., m., bacon. E»S

beijman (BEICK-man), n., m., basement. "Tío Raúl has a huge storage en el beijman." {CA} E»S

beis (BEIS), n., m., baseball. E»S Also BEISBOL.

beisbol (BEIS-ball), n., m., baseball. E»S Also BES.

beisbolista (beis-bo-LEES-ta), n., m/f., person who plays baseball.

beismen (BEIS-men), n., m., 1. basement. Also LIMINAL. 2. n., m., male baseball players. "Ellos son las beismen del eguipo."

bejaviorismo (bee-chai-bohr-EES-mo), n., m., behaviorism. "He loves la ciencia del bejaviorismo." E»S

belduque (bel-DOO-ke), n., m., knife. "He committed the crime with a belduque." {Ch}

benchmarcar (bentch-mar-KAR), v., to sit on the bench athletically. E»S

benchmark (bench-MARK), n., m/f., 1. inactive athlete. 2. one who perpetually 'sits the bench' in athletic events. E»S

benchmarquin (bench-MAR-keen), adj., m/f., inactive athletically. "This season él está benchmarquin." E»S

bendéi (ben-DE), n., m., band-aid. {Ch} E»S

bendito (ben-DEE-toh), n., m., outlaw. In Sp. *bendito* means holy. S»E Also BANDIDO.

benkenpura (ben-ken-POO-ah), n., f., baking powder. "She baked the cake with el benkenpura." {Ch} E»S

berrego (boh-RRE-go), adj., bothersome. "Stop being a berrego!" {Ch}

beseler (bes-SEH-ler), n., m., best-seller. E»S

betabelero (ba-the-be-LEH-roh), n., m., harvester of sugar beets. {Ch}

betnic (BET-neek), adj., m/f., related to the beatnik style. "I remember el movimiento betnic of the 1960s."

bias (BEE-as), n., m., bias. "He didn't get el trabajo porque había bias." Sp. *parcialidad*. E»S

bica (BEE-ka), n., f., money. Also BICOCA {Ch}

bicarbonato (bee-kar-bo-NA-toh), n., m., bicarbonate. E»S {Ch}

biciar (bee-SEE-ar), n., m., V.C.R. "Ahora que Chucho me enseñó cómo usar el biciar, I never miss a soap opera episode." E»S

bicoca (bee-KO-ka), n., f., money. {Ch} Also BICA.

bicyclo (bee-SEE-clo), n., f., bicycle. E»S

bidanoffer (beed·an·OH·fer), exp., purchase price. Eng. *bid and offer.* "Tenemos el bidanoffer and we're buying the house." E»S

bife (BEE·fe), n., m., beef stake. Also BIFSTAIK.

bigbán (big·BAN), n., m., the Big Bang. "Do you believe en Dios o en el bigbán?" E»S

bigsho (BIG·shot), n., m., big shot. E»S

bil (BEEL), n., m., bill. E»S

bildin (BELL·din), n., m., building. {CA, NR} E»S

biles (BEE·less), n., m., bills. "Don't forget de pagar los biles. Don't forget to pay the bills." E»S Also BILL.

binocular (bee·noh·koo·LAR), n., m., binoculars. {Ch}

bipear (BEE·pear), v. to page someone, to use a beeper. "Por favor bipéae a tu padre to know at what time he'll return." E»S Also BIPIAR.

bíper (BEE·per), n., m., phone pager. E»S

bipiar (BEE·pear), v. to page someone, to use a beeper. "Por favor bipíale a tu padre to know at what time he'll return." S»E Also BIPEAR.

bipop (BEE·pop), n., m., the jazz movement of bebop. E»S

bironga (bee·RON·ga), n., f., beer. {Ch}

bironguero (bee·ron·GUE·ro), n., m/f., beer drinker. {Ch}

birria (bee·REAH), n., f., beer. In Mex. Sp., a type of beef meat. {Ch}

bisi (BEE·see), adj. busy, in a rush, stressed out. "Mami, deja llamarte pa' tras que estoy super bisi y el jefe está a llegar del lonche." {CA} E»S

bisnes (BEES·ness), exp., in good shape. "Estoy en bisnes, so I'm in fine shape." E»S

bisnesplan (bees-ness-PLAN), n., m., business plan. E»S

bitoque (bee-TO-ke), n., m., hose nozzle. "Apreta el bitoque of the horse." {Ch}

blaf (BLAPH), n., m/f., bluff. {Ch} E»S

blakout (blak-OUT), n., m., 1. black-out; 2. blockage. "Hay blakout por todas partes del pueblo because we are without electricity." E»S

blanco (BLAN-koo), n., m., 1. blank. "Please, llene el blanco." {CS} 2. white person. S»E

blanquillo (blan-KEE-looh), n., m., testicle. "They hit him en los blanquillos." {Ch, Mex} S»E

blessin (BLE-zeen), n., m., blessing, benediction. E»S

blichar (BLEE-tchar), v., to bleach. {Ch} E»S

blichi (BLEE-tchee), n., m., bleach. "Ella usó el blichi para lavar the cloths." {Ch} E»S

blister (BLEES-ter), n., m., 1., blister. "The shoes are tight. Me sacaron un blister. Sp. *ampolla*. 2. blister pack. E»S

blizar (BLEE-zar), n., m., blizzard. E»S

bloaut (bloh-OUT), n., m., 1. blowout merchandise sale. "Fuiste al bloaut sale de Macy's?" {Ch} 2. Electric blowout. "Había tremendo bloaut en la ciudad when the power went off." E»S

blodimari (blo-ddee-MEH-ree), n., m., Bloody Mary. Mixed alcoholic beverage. E»S

blof (BLUF), n., m., bluff. E»S

blofeador (bloh-phea-DOOR), n., m/f., person that bluffs. "Es tremedo blofeador at school." {Ch}

blofear (bloh-PHEAR), v., to bluff. E»S

bloque (BLOH-kee), n., m., city block. E»S

bloquear (BLOO-kear), v., to block. E»S

bloqueo (bloh-KEE-oh) n., m., embargo. "Si no existiera el bloqueo in Cuba, we could buy zapatos buenos."

bluchip (BLUE-cheep), n., m., blue chip. "Quédate con el estok de bluchip." {SCS}

blumers (BLOO-mers), n., m., panties. {S} S»E

blus (BLOOS), n., m., blues, musical rhythm. E»S

bluyins (bloo-JEENS), n., m., blue jeans. E»S

bob (BOB), n., m., bobsled. E»S

bobear (BOH-bear), v., to drink. {Ch}

bobo (BOH-boh), n., m., a drunk. {Ch}

bobteil (bob-TAIL), n., m., a type of dog. E»S

bochinche (boh-CEEN-cheh), n., f., melee. {Ch}

body (boh-DEE), n., m., 1. body of a dress. "Le queda bien el body del vestido." 2. bodyboarding. "Voy a hacer body. I'm going body-boardín." {CA} E»S

bodybildin (bo-dee-BEEL-deein), n., m., body building. E»S

bodybord (bo-dee-BORD), n., m., bodysurfing board. E»S

boicot (BOY-kot), n., m., boycott. E»S

boicotear (boy-koh-TEHAR), v., to boycott. {Ch} E»S

boicotero (boy-koh-THE-roh), n., m/f., boycotting person. E»S

boila (BOY-lah), n., f., heating appliance, boiler. "Si Bebo no acaba de arreglar la boila, the five of us will freeze here." {NY} E»S

boiletals (boy-LEH-tals), exp., "Hot damn!" {Ch}

boiscau (BOY-scauh), n., m., boy scout. {CA, NR} E»S

bolardo (bo-LAR-doh), n., m., bollard. {S}

bolero (boh‑LEH‑roh), n., m., type of Spanish dance. The term dates from 1787. S»E

bolevar (boh‑leh‑VAR), v., to dance. "Jael bolevea en el Salón Xalapa." {Ch}

bolillo (boh‑LEE‑yoh), n., m., foreigner. Mex. Sp. *bolillo* is a type of bread. {Ch} Also GRINGO.

bolina (boh‑LEE‑nah), n., f., bowline. "Navegar de bolina." E»S

bolo (BOH‑loh), n., m., dollar. {Ch}

bolsudo (bol‑SOO‑doh), n., m., full of money. {Ch}

bom (BOM), n., m., homeless person. Eng. *bum*. "You must graduate on time si no quieres acabar como un bom en el *dauntaun*." {CA} E»S

bomper (bom‑PER), n., m., car bumper. E»S

bonanza (boh‑NAN‑zah), n., f., 1. fair ocean weather. Applied to a profitable enterprise. {SW} 2. unexpected benefit. The term dates from 1842. S»E

bonche (BON‑tche), n., m., 1. bunch, handful. 2. joking bunch. Sp. *relajo*. {CA} S»E Also BONCHINCHE.

bonchinche (bon‑TCHEEN‑tche), n., m., joking bunch. {CA} Also BONCHE.

bondo (BON‑dl), n., m., bundle. "Se encogió de dolor como un bondo." E»S

bonque (BON‑keh), n., m., bunk. "Los hermanos tienen un bonque de dos camas." {Ch} E»S

bonquear (bon‑KEAR), v., to sleep. "Juan se bonquea a las 7 PM." {Ch} S»E Also BONQUIAR.

bonquiar (bon‑KEAR), v., to sleep. "Juan se bonquía a las 7 PM." {Ch} S»E Also BONQUEAR.

boom (BOOM), n., m., explosion. E»S

bookaroo (BO‑ko‑roo), n., m., cowboy. From Sp. *vaquero*. {SW} S»E

boquineta (bah‑kee‑NEH‑tah), n., m., left palate. {Ch}

borachio (boh‑RRA‑tchio), n., m., drunkard. The term dates from the early 17ᵗʰ century.

bordera (bor‑DEH‑rah), n., f., boarding house owner. "Si deseas una noche, talk to the bordera." {Ch}

borderígena (bor‑deh‑REE‑ge‑nah), n., m/f., border citizen.

borderline (bor‑der‑LAIN), n., m/f., 1. used to describe one of constant mental instability: "schizoid." E»S

bordin (BOOR‑deen), v., to board. E»S

borguaize (BOR‑wai‑zer), n., f., beer. After brand name. "On July 4ᵗʰ bebemos borguaizes." {CA} E»S

boricua (boh‑REE‑kua), n., m/f., person from Puerto Rico. The term dates from 1887. Also BORINCANO, BORINQUEÑO, PUERTORRIQUEÑO and PORTORRIQUEÑO.

borin (BOH‑reen), exp., boring. "Este libro de Carlos Fuentes está borin." E»S

borincano (boo‑reen‑KAH‑no), adj., m/f., person from Puerto Rico. Also BORICUA, BORINQUEÑO, PUERTORRIQUEÑO and PORTORRIQUEÑO.

Borinquén (bo‑reen‑QUEN), n., original name of Puerto Rico. Also spelled without the accent.

borinqueño (bo‑rin‑QUE‑nyo), n., m/f., person from Puerto Rico. The term dates from 1647. {NR, PR} Also BORICUA, BORIN‑ CANO, PUERTORRIQUEÑO and PORTORRIQUEÑO.

borol (boh‑ROL), n., f., bottle. "Dame el borol de milk, por favor." {Ch} E»S

boroleado (boh‑ro‑LEA‑do), adj., bottled. {Ch} E»S

bos (BOS), n., m/f., 1. chief person. 2. bus. "Se subió en el bos." E»S Also BUS.

bosal (bo-SAAL), n., m., headstall for a horse. Sp. *bozal.* S»E

boslain (bos-LAIN), n., f., public transportation route, bus line. "Coge el boslain número 45." {Ch} E»S

bote (boh-THE), n., m., prison. "María cometió un crimen y terminó en el bote." {Ch, Mex}

botombrá (bo-tohn-BRA), n., m., 1. Wonderbra. 2. push-up bra. "Ella usa that type of botombrá." E»S

botoniar (boh-toh-NEAR), v., buttoning process. "Mam, can you help me a botoniarme el vestido? E»S

boulin (BOW-LEEN), n., m., bowling. E»S

boupepo (BOW-pee-po), n., m., boat people. Used in plural. "Ellos llegaron a Cayo Hueso como boupepo." {CA} E»S

box (BOKS), n., f., 1. box. Sp. *caja.* 2. ticket booth. In Sp. *taquilla* is box office. "He left los boletos en la box." E»S

boxeador (boh-XEA-door), n., m., boxer. E»S

boxear (boh-XIAR), v., to box, athletic activity. E»S

braca (BRA-kah), n., f., 1. school break. "I'm de braca." 2. automobile brake. "I stop con la braca." Also BRECA. {Ch} E»S

brainstormin (BREIN-stor-meen), v., to think profusely. "Estoy brainstormin ahora, please don't bother me." E»S

braun (BRAWN), exp., brown. "Nosotros somos los brauns de la clase." {SW} E»S

braunsero (brown-ZEH-ro), n., m., specialist in suntan. {Ch} E»S

braunsiar (brown-ZEEAR), v., 1. to darken one's skin. 2. to suntan. {Ch} E»S

bravado (bra·VA·doh), n., m., 1. arrogance. 2. bluster. 3. affectation of heartiness. "Y ése, qué bravado!"

bravo (BRA·voh), exp., approval or appreciation.

break (BREIK), n., m., vacation. E»S Also BREQUE.

breakdown (BREIK·dawn), n., m., collapse. E»S

breca (BREH·kah), n., f., automobile brake. "Step on the brekas." Also BRACA. {Ch}

briche (BREE·tche), n., m., bridge. {Ch} E»S

brifin (BREE·fin), n., m., to update, to brief. "Did you give toda la información en el brifin?" E»S

bro (BRO), n., m., friend. {Ch} Also BATO, BRODER, BROTHER, GÜE, GÜEY, and VATO.

broder (BRO·ther), n., m., friend. Eng. *brother.* E»S Also BATO, BRO, BROTHER, GÜE, GÜEY, and VATO.

broker (BRO·ker), n., m., 1. financial and real·estate advisor. 2. intermediary. Also BROQUER. E»S

bronca (BRON·kah), n., f., trouble. "Hey, fool, quieres bronca?" {Ch, Mex}

bronch (BRONCH), n., m., brunch. "El bonch es a las 11 AM en casa de Leticia." E»S

bronco (BRON·kooh), n., m., untamed horse. The term dates from 1866. S»E

bróquer (BRO·ker), n., m., 1. financial and real·estate advisor. 2. intermediary. E»S Also BROQUER.

broquerage (BRO·ke·rah·ge), n., m., financial arrangement. Eng. *brokerage.* E»S

brother (BRO·der), n., m., friend. {Ch} S»E Also BATO, BRO, BRODER, GÜE, GÜEY, and VATO.

buca (BOO·kah), n., f., girl. "Esa buca está hermosa." {Ch}

buche (BOO·tche), n., m., ass. {Ch}

buchería (boo·tche·REE·ah), n., f., butcher store. "Dad went to la buchería to buy meat." E»S

buckaroo (boo·kah·ROO), n., m., cowboy. Sp. *vaquero.* S»E

budin (boo·DEEN), n., m., pudding. S»E

buffer (BOOh·pher), n., m., intermediary. E»S

bug (BOG), n., m., computer malfunction. "La computadora tiene un bug." {CS} Also BUGO. E»S

buget (BOO·jet), n., m., budget. E»S

bugo (BOO·goh), n., m., bug. {CS} E»S Also BUG.

bugui-bugui (BOO·gee·BOO·gee), n., m., boogie woogie. E»S

búlava (BOO·la·vah), n., m., boulevard. Fr. *boulevard.* {Ch}

bulchiteador (bool·tchee·TEHA·door), n., m., person that offends, bullshits. E»S

bulchitear (bool·tchee·TEHAR), v., to bullshit, to offend. "Todo el día that guy bulchitea." {Ch} E»S

bulchiteo (bool·tchee·THE·oh), n., m., action of offending. Eng. *bullshitting.* "El bulchiteo de la gente at the stadium is unbearable." E»S

buldog (BULL·dog), n., m., police officer. {SW and Mex} E»S

bule (BOO·leh), n., m., bully. {Ch} E»S

bulto (BOOL·toh), n., m., ghost. {Ch}

bumeran (boo·meh·RAN), n., m., boomerang. E»S

búngalo (BOON·ga·loh), n., m., bungalow.

bucreo (boo·KEH·roh), n., m., person booking travel and other arrange-ments. "El bucreo made our arreglo to Buenos Aires." E»S

buqui (BOO·kee), n., m., child. "Ella tiene un buqui de 10 años." {Ch}

buquiar (boo-KEEHAR), v., 1. to book. 2. to make plans. 3. to arrange. "Buquiaste el show de Gloria Estefan?" E»S

burguesa (bur-GUE-sah), n., f., hamburger. E»S

burlista (boor-LEES-tah), n., m/f., joker. {Ch}

burnishear (boor-nee-SHEHAR), v., to burnish. E»S Also BURNI-SHIAR.

burnishiar (boor-nee-SHEHAR), v., to burnish. E»S Also BURNI-SHEAR.

buró (boo-ROH), n., m., 1. chest, furniture. 2. central office. From Frn. *bureau.*

burrito (boo-REE-to), n., m., large taco wrap made of letucce, beans, rice and chicken or beef. Burritos are unknown in Mexico. In Sp. *burrito* means little donkey.

burro (BOO-roh) n., m., 1. lazy. Mexicanism. 2. donkey. The term dates from 1800. S»E

burro trail (boo-rroh TRAIL), n., m., a trail made of repeated donkey trampings.

bus (BOS), n., m/f., 1. chief person. 2. bus. "Se subió en el bus." E»S Also BOS.

buscatoques (boos-ka-TO-kes), n., m., drug addict. {Ch}

butear (boo-TEHAR), v., to boot. {CS} E»S

butifulpeepo (boo-tee-foo-PEE-po), n., m., 1. group of rich, famous and exciting people. 2. jetset. "I was told que you anda con los butifulpeepo." E»S

byte (BAI-teh), n., m., computer term. {CS} E»S Also BAITE.

C

caballada (kah‑ba‑YAH‑dah), n., f., a band of saddle horses. S»E

caballero (kah‑ba‑YEH‑ro) n., m., gentleman. S»E

caballo (kah‑BA‑yo) n., m., 1. horse. 2. heroin. "Judith es adicta al ca‑
ballo." {Ch}

cabana (ka‑BA‑nah), n., f., rustic habitation. Sp. *cabaña.*

cabbagio perfumo (ka‑ba‑yo per‑PHOO‑moh), n., m., 1. cheap, rank
cigar. 2. jocular. The term dates from the late 19th century.

cabestro (ka‑BES‑tro), n., m., horse‑hair rope halter.

cabina (ka‑BEE‑nah), n., m., cabin. E»S

cabrón (ka‑BRON), n., m., son of a bitch.

caburro (ka‑BOO‑rroh), n., m., cowboy. {Ch}

cabus (ka‑BOOS), n., m., caboose. {Ch} S»E

cacafuego (ka‑ka‑PHOOEH‑go), n., m., spitfire, braggart, bully. The
term dates from the late 17th to early 18th centuries.

cach (KATCH), n., m., cash. E»S

cachar (KA·tchar), 1. to catch, to gore, to butt. 2. to copulate. 3. To check out {Ch}. "Huevón, cacha! Check it out, man!" E»S Also QUECHAR and KECHAR.

cacher (KA·tcher), n., m., baseball position. E»S

cachu (KA·tchoo), n., m., ketchup. "Pásame el cachu, please!" E»S

cadi (KA·dee), n., m., golf caddie. E»S

café (ka·PHEH), n., m., small restaurant. "We met at a café for coffee." Sp. *café*. S»E

cafishio (ka·FEE·shyo), n., m/f., 1. person of Creole descent. 2. *compadre*. Also CANFINFLERO. 3. professional pimp. {Arg}

caite (KAI·teh), n., f., kite {Ch} E»S

calaboose (ka·la·BOOZ), n., f., jail. The term dates from 1837–40. Sp. *calabozo*. {Ch} S»E Also CALABOS.

calabos (ka·la·BOZ), n., f., jail. The term dates from 1837–40. Sp. *calabozo*. {Ch} Also CALABOOSE.

calco (KAL·ko), n., m., shoes. {Ch}

caleraidi (ka·ler·AY·dee), n., f., caller ID, telephone device. {CS} E»S

calgüeitín (kal·wai·TEEN), n., m., call·waiting, phone device. E»S Also CALWAITIN

caliente (ka·LEEHN·teh), n., m., heating system. "It's cold outside y no sirve el caliente." {NY}

Califa (ca·LEE·fa), adj., resident of California of Mexican descent. "They are Califas from L.A., sabes?" {SW} Also CALIFEÑO and CALIFORNIO.

Califeño (kah·lee·PHE·nyo), adj., m/f., resident of California of Mexican descent. "They are Califeños from L.A., sabes?" {SW} Also CALIFA and CALIFORNIO.

Californio (ka-lee-FOR-nio), adj., resident of California of Mexican descent. "They are Californios from L.A., sabes?" {SW} Also CALIFA and CALIFEÑO.

call (KAL), n., f., buying option. "Ella compró una call." S»E

Callao painter (ka-YAOH pain-ter), n., m. evil-smelling gas arising from sea at that port: nautical coll. The term dates from the late 19th and early 20th centuries.

Callao routine (ka-YAOH Rooh-TEEN), n., m., discipline that is free and easy. Also CALLAO SHIP.

Callao ship (ka-YAOH sheep), n., m., free and easy life, Grainville. At Callao, the principal seaport of Peru, things seem, to a naval rating, to be free and easy. Also CALLAO ROUTINE.

calle (KA-yeh), n., f., 1. street. "He lives in Calle Ocho." 2. cloak. 3. gown. The terms dates from 1670.

caló (ka-LOH), n., m., slang. "You got to know caló." {Ch}

calwaitín (kal-wai-TEEN), n., m., call-waiting, phone device. E»S Also CALGUEITIN.

calzonera (kal-zo-NEH-ra), n., f., 1. Mexican-style pants with buttons down the sides, worn by early traders. 2. low-life people. Exp. used to ridicule. "The gente of the countryside is calzonera!" {Ch}

cam (KAM) 1. camera. 2. camisole. Also CAMMY and EXAMI.

cama (KA-ma), n., f., bedroll.

camcorder (KAM-kor-der),n., m., hand-held video camera. E»S

cameador (ka-mea-DOOR), n., m/f., hard worker. {Ch}

camera oscura (KA-me-rah os-KOO-rah), n., f., 1. dark room. 2. adj., facetious.

cameraman (ka-me-rah-MAN), n., m., camera person. E»S

camesa (ka-ME-sa), n., f., 1. shift. The term dates from 1660. 2. shirt. Sp. *camisa.* 3. commission.

camino real (ka-mee-no REEAL), n., m., a main highway.

cammy (KA-mee), n., f., camisole. E»S Also CAM and EXAMI.

campanyero (kam-pa-NYEH-roh), n., m/f., comrade. Sp. *compañero.*

campear (kam-PEAR), v., to camp. E»S

campin (KAM-peen), n., m., campground. "We look por el campin." E»S

campión (kam-PEEON), n., m., champion. "El fue el campion de la competencia." Sp. *campeón.* E»S

campo (KAM-poh), n., m., 1. military camp. "John está en el campo militar número ochenta y tres." 2. playground, field: schools. 3. rural section. The term dates from the 1890s.

canfinflero (kan-pheen-PHLEH-roh), n., m/f., 1. person of Creole descent. 2. *compadre.* Also CAFISHIO.

canochi (ka-NOH-tchi), n., m., a Mexican person still living south of the Rio Grande. {Ch}

canó (ka-NO), n., m., canoe. "Usamos el canó en el Río Uzumacinta." Sp. *canoa.* {Ch}

canon (KA-non), n., m., narrow valley between high cliffs. Eng. *canyon.* The term dates from 1834. S»E

canquiza (kan-KEE-za), n., f., beating. "He got una buena canquiza." {Ch}

cantina (kan-TEE-nah), n., bar, saloon. S»E

cantinflada (kan-teen-FLEA-dah), n., f., nonsense talk. {Ch}

cantinflear (kan-teen-FLEHAR), v., to speak nonsense. {Ch}

cantinflesco (kan-teen-FLES-ko), adj., attribute of cantinflada. {Ch}

canton (KAN·ton), n., m., house. "Could we stop by my canton so I can grab my jacket?" {ELA}

capa (KA·pah), n., f., baseball cap. "Check out my new capa."

capangaun (kap·an·GOON), n., m., cap and gown. "When you graduate, yo te compraré un capangaun." E»S

caponera (ka·poh·NEH·ra), n., f., herd of geldings.

caporal (ka·poh·RAL), n., m., boss.

capote (ka·POH·te), n., m., coat. "Nice capote, hombre!"

caqui (KA·kee), n., m., khaki color. "He wears los pantalones caqui." E»S

caracter (ka·RAC·ter), n., m., fictional personage. "Me gusta el caracter de Sancho Panza." E»S

carajo (ka·RAH·kcho), n., m., ox driver, mule skinner, or other base character. Sp. *carajo* is exp. for goddam. "No entiendo ni un carajo." {Ch, Mex, and Cen Am} Also CARAMBA.

caramba (ka·RAM·ba), exp., goddam. Also CARAJO. {CA, NY}

caravanin (ka·ra·BA·neen), n., m., travel by land. "Ese verano fui en un caravanin al Grand Canyon." S»E

caravansera (ka·ra·van·SEE·ra), n., f., 1. railway station: Orig. ca. 1845–1900. 2. boxing fight.

carga (KAR·gah), n., f., load of narcotics. "Cogiste la carga of cocaine?" {Ch}

cargo (KAR·goh), n., m., 1. a ship's load. "The cargo ship brought steel from Russia." 2. contempt for a person. "He has a cargo against Dolores for not being honest." The term dates from 1657.

carita (ka·REE·tah), exp., tease. "Qué carita eres! What a tease you are!" {Ch}

carnal (kar·NAL), n., m., friend. {Ch} Also BATO, BRO, BROTHER, GUE, GUEY, and VATO.

carnalismo (kar·na·LEES·mo), n., m., camaraderie. {Ch}

carne asada (KAR·neh a·ZAH·dah), n., f., 1. barbecued meat. 2. in trouble. "Martin committed a crime. Es carne asada!" {SW}

carne seco (KAR·ne ZE·koh), n., m., jerky. "Javier es un estúpido. Es carne seco." In Sp. *carne seca* means dry meat. {CA}

carpeta (kar·PE·tah), n., f., carpet. Sp. *alfombra*. E»S

carretero (ka·rre·TE·ro), n., m., wagoneer.

carrier (KA·rrier), n., m., high·frequency electric radio. E»S

carro (KA·rroh), n., m., automobile. E»S {Ch, Mex}

carrucho (ka·RROO·tcho), n., m., automobile. "That is a real nice car·rucho." {ELA} Also RANFLA.

carrunfla (ka·RRUN·fu·lah), n., f., jalopy. {Ch}

carta (KAR·ta), n., f., cart. "En el supermercado usamos la carta." In Sp. *carta* means letter. E»S

carta de marca (KAR·ta de MAR·kah), n., f., trademark. E»S

carton (kar·TON), n., m., cigar carton. "Compró un cartón de Marlboro." E»S

cartún (kar·TOON), n., m., cartoon. "Los niños watched los cartúns esta mañana." E»S

casa (KA·sah), n., f., 1. cab. 2. a writ of capias. The term dates from 1650.

caset (KA·set), n., m., cassette. E»S

cash (KASH), n., m., liquid money. E»S

cashancary (KASH·an·ka·ree), n., m., self·service. "No me gustan las bombillas de cashancarry en la gas stations." E»S

cashflo (KASH-flo), n., m., cash flow. "Qué tal el cashflo? Do you need money?" E»S

caso (KA-so), n., m., to take a room or a flat and become a prostitute. "She only married him for the money, moved in and became a caso." Also GOCASO.

cast (KAST), n., m., cast of actors. E»S

casteyanqui (kas-te-YAN-kee), n., m., mestizo language, part English, part Spanish, used predominantly in the United States since WWII. Also GRINGONOL and SPANGLISH.

castin (KAS-teen), n., m., talent casting. E»S

castumar (cas-TOO-mar), n., m/f., customer. {NY} E»S

casual (KA-sual), adj., informal in dress style. "Juanita dresses muy casual."

casualidad (ka-sua-lee-DAD), n., f., casualty. "La cantidad de casualidades en 9/11 fue de aproximadamente 3,000 personas." E»S

catamarán (ka-ta-ma-RAN), n., m., athletic boat. E»S

cauboy (kau-BOY), n., m., vaquero, cowboy.

cauch (KAUTCH), n., m., couch. E»S

caucioso (kau-SEEHO-so), adj., m., f., cautious.

caudillo (kau-DEE-llo), n., m., chief of state. The term dates from 1852. S»E

caverango (ka-ve-RAN-go), n., m., wrangler. Sp. *caballerango.*

Cayo Hueso (KAH-yo-GUE-so), n., Key West, Florida. {CA}

cazatalentos (ka-za-ta-LEN-tos), n., m., 1. talent scout. 2. headhunter.

cedecé (se-DE-ce), n., m., acronym for *casa del carajo,* distant place. "Better that we go to Paco's party in the Gables because for real Luli lives in cedecé." {C}

cederom (se-de-ROM), n., m., CD-ROM. {CS} E»S

cefeó (ze-phe-O), n., m., Chief Financial Officer, financial administrator. "He is the cefeó in an important corporation."

ceis (KEIS), n., m., case. 1. legal case. "El abogado works en el ceis de Juanito." 2. exp. in case. "Estaré en casa ceis you arrive early." {Ch} E»S Also KEIS and QUEIS.

celiula (ZE-liu-lah), n., f., cellular phone. E»S

celiular (ze-liu-LAR), v., to use the cellular phone, to call through a cellular connection.

centavo (zen-TA-vo), n., m., 1. cent. 2. coin. The term dates from 1883.

ceo (ZE-oh), n., m., Chief Executive Officer. Sp. *gerente general.* E»S

ceoó (ze-oh-O), n., m/f., Chief Operating Officer. Sp. *gerente de operaciones.* "She got the posición de ceoó en Sears." E»S

cero (ZE-roh), n., m., kindergarden. "My son está en cero." {Ch}

chacketa (tcha-KE-tah), n., f., jacket. Eng. *jacket.* E»S Also CHAQUETA.

chacón (tcha-KON), n., m., shotgun. {D}

chainear (TCHAI-near), v., to shine, to polish.

chale (TCHA-leh), exp. of disagreement. Also used for *goddamit.* "Chale, why did you have to blow up my spot like that?" {ELA}

chamba (TCHAM-ba), n., f., job. "I'm looking for chamba in business." {Ch}

champeón (TCHAM-peon), n., m/f., champion. E»S Also CHAMPION.

champión (TCHAM-peeon), n., m/f., champion. E»S Also CHAMPEON.

champú (TCHAM-poo), n., m., shampoo. E»S

chance (TCHAN-ze), n., m., chance. "El tuvo el chance de ganar the prize but lost it." E»S Also CHANSA

chankli (TCHAN‑kee), n., f., 1. individual of low class. 2. person of ill repute. From Cuban *chancletera*. The origin is probably the loud, wooden sandals (*chancletas de palo*) worn by poor women. "Mayra tries to act as a princess but she's chankli." {CA}

chanks (TCHANX), n., f., sandals, flip‑flops. Sp. *chancletas*. "La arena está que quema and I left my chanks in the room." S»E

chansa (TCHAN‑za), n., f., chance. E»S Also CHANCE.

chante (TCHAN‑teh), n., m., home. "My chante is comfortable." {ELA}

chaparral (tcha‑pa‑RRAL), n., m., region characterized by shrubs and low, evergreen oaks. The term dates from 1850. S»E

chaparro (tcha‑PA‑rro), n., m., low, evergreen oak. S»E

chaps (TCHAPS), n., f., leather overalls or leg protectors. Sp. *chaparejos*. {CA}

chaqueta (tcha‑KE‑tah), n., f., jacket. Eng. *jacket*. E»S Also CHACKETA.

charco (TCHAR‑ko), n., m., divide. {Mex}

charqui (TCHAR‑kee) n., m., beef jerky.

chat (TCHAT) n., m., talk, dialogue. {CS} E»S

chateante (tcha‑TEAN‑te) n., m/f., 1. talker. 2. n., m., idle talk. Sp. *cháchara*. {CS} E»S

chatear (tcha‑TEAR), v., to chat. E»S

chateo (tcha‑TEAN‑do), n., m., act of chatting. {CS} E»S

chauer (TCHAW‑ser), n., m., 1. bathroom shower. 2. baby shower. "Ella went al chauer del bebé." 3. wedding shower. E»S

checar (tche‑KAR), v., 1. to check. 2. to make oneself present. 3. exp., See you later! "Hay te checo!" E»S Also CHEKIAR and CHEQUEAR

chei (TCHEI), n., f., Venetian blinds. From Eng. *shades*. E»S

cheirman (TCHEIR-man), n., m., chairman. E»S

cheirwoman (tcheir-WO-man), n., f., chairwoman. E»S

chekiar (tche-KEEAR), v., 1. to check. 2. to make oneself present. 3. exp., See you later! "Ay te chekio!" E»S Also CHECAR and CHE-QUEAR.

chekirau (tche-kee-RAU), exp., Check it out! E»S

chekup (tche-KUP), n., m., check-up. "Raul, your chekup con el médico is a las nueve." E»S

chela (TCHE-la), n., f., beer. "Tomamos una chela?"

cheque (TCHE-ke), n., business check. "Send cheque or dinero." E»S

chequear (TCHE-ke-ar), v., 1. to check. 2. to make oneself present. 3. exp. See you later! "Hay que chequear el equipaje." E»S Also CHECAR and CHEKIAR.

chequeo (tche-KE-o), n., m., the act of checking out. "Nos cheque-amos later on."

cherife (tche-REE-phe), n., m., sheriff. E»S

Chicanadian (tchee-ka-NE-dian), n., m/f., Mexican from Canada. {Ch}

Chicanglo (tchee-KAN-glo), n., m/f., Anglicized Chicano. {Ch}

Chicano (TCHI-ka-no), adj. and n., m/f., person of Mexican background living in the United States. The etymology is believed to be from *mexicano*. The term dates from 1947.

chick (CHEEK), n., m/f., 1. friend. {SW} Sp. *chico*. 2. n., f., attractive woman. S»E Also CHICKO and CHICO

chicko (CHEE-koh), adj. and n., m., young man, average guy. "She dated a chicko from another town." In Sp. *chico*. {SW} Also CHICK and CHICO.

chico (CHEE·koh), adj., m., young man, average guy. S»E Also CHICK and CHICKO.

Chihuahua (tchee·GUA·gua), n., m., variety of small dog named for the region in Mexico, the state of Chihuahua, from which it originates. S»E

chile (CHEE·leh), n., m., American Indian chili. S»E

chile relleno (CHEE·le re·YEH·noh), n., m., Mexican·type dish made with cheese, meat, and chile pepper.

chilear (TCHEE·lear), v., to chill out. "Después del juego, se duchó para chilearse." E»S Also CHILIAR.

chili (TCHEE·lee), n., m., hot pepper. The term dates from 1662. S»E

chiliar (TCHEE·lee·ar), v., to chill out. "Después del juego, se duchó para chiliarse." E»S Also CHILEAR.

chili·con·carne (CHEE·le kon·KAR·neh), n., m., Tex·Mex dish made with beans, shredded meat, and tomato sauce. S»E

chinchilla (tcheen·TCHEE·ya), n., f., 1. a type of rodent. Usage dates from 1604. 2. nuisance. "Don't be una chinchilla."

chinchín (tcheen·tcheen), n., m., a toast. From the sound of clashing wine cups. "Hagamos chinchín. Let's make a toast."

chintz (CHEEN·tz), n., m., blue Jeans. E»S

chip (CHEEP), n., m., computer chip. {CS} E»S

chipe (CHEE·peh), adj., m/f., cheap. "La nueva TV was muy chipe!" E»S

chips (CHEEPS), n., m., potato snacks. {Mex, SW} E»S

chiriando (CHEE·rehan·doh), v., unethical practice of using slick means to obtain something. "Nay fool, that carnal estaba chiriando." Eng. *cheating*. E»S

chirlider (CHEER·lee·dehr), n., m/f., cheerleader. E»S

chirro (CHEE-rroh), n., m., sheet rock, the material used for thin walls in modern Miami homes. "Cuando le convirtieron el garage en otro cuarto they put paredes de chirro." {CA}

chivaldría (tchee-val-DREE-a), n., f., chivalry. "Don Quixote era lec-tor de novelas de chivaldría." E»S

chiz (TCHEEZ), n., m., gossip. "A su tía le gusta el chiz." Sp. *chisme.* {CA}

cho (TCHOO), n., m., refers to an embarrassing moment, usually in public. Eng. *show.* It is the equivalent of *making a scene.* "No hagas un cho cuando no es pa' tanto." {CA} E»S

choke (TCHOK), n., m., classroom chalk. E»S Also CHOQUE.

chokeador (tcho-kea-DOOR), n., m/f., person given to shocks. E»S

chokear (tcho-KEAR), v., to shock. E»S

cholla (TCHO-yah), n., f., a spiny variety of cactus. {Mex}

chomba (TCHOM-ba), n., f., jumper. {NY} Also CHOMPA.

chompa (TCHOM-pa), n., f., jumper. {NY} Also CHOMBA.

chopear (tcho-PEHAR), v., 1. to chop. 2. to buy, to shop. "Vamos a chopear en el *mol.*"

chopin (TCHO-peen), n., m., 1. Shopping center, mall. "En el chopin van a poner un Pollo Tropical." 2. v., the act of going shopping. "Enseguida que cobra se va chopin a las tiendas."

choque (TCHO-ke), n., m., 1. classroom chock. Also CHOKE. 2. state of shock and nervous depression. "El broder sufrió un choque muy fuerte after the accident." E»S

choqueado (tche-KEA-do), adj., m/f., surprised, dumbfounded. Eng. *shocked.* E»S Also CHOQUE, CHOKEAR.

chor (TCHOR), n., m., endeavor. E»S

chors (TCHORS), n., m., short pants. "Necesitas some chors to swim." E»S

chusar (tchoo-ZAR), v., to choose. E»S Also CHUSIAR.

chusiar (tchoo-ZEEAR), v., to choose. E»S Also CHUSAR.

chut (TCHOOT), n., m., strong kick. Soccer term. E»S

chutar (tchoo-TAR), v., to shoot. "Juanito chuta un golazo." E»S Also CHUTEAR.

chutear (tchoo-TEAR), v., to shoot. Soccer term. E»S Also CHUTAR.

cibberpunc (zee-ber-PUNK), n., m/f., cyberpunk. {CS} E»S

ciber (ZEE-ber), n., f., cybernetic. {CS} E»S

ciberespacio (zee-ber-es-PAH-zeeo), n., m., cyberspace. Also spelled *cyberespacio.* {CS} E»S

cibernauta (zee-ber-NAU-tah), n., m/f., an internet navigator. {CS} E»S

cibernético (zee-ber-NEH-tee-ko), n., m., that which pertains to cybernetics. {CS} Also spelled *cybernético.* E»S

ciberparque (zee-ber-PAR-keh), n., m., cyberpark. {CS} E»S

ciberpunc (zee-ber-PUNK), n., m/f., cyberpunk. Also spelled *cyberpunc.* {CS} E»S

ciberspanglish (zee-ber-SPAN-gleesh), n., m., Spanglish terms used on the Internet. Also spelled *Cyber-Spanglish.* E»S

cibervato (zee-ber-BAH-to), n., m., Chicano internet user. E»S

cibolero (zee-bo-LE-roh), n., m., buffalo hunter. From Sp. *cíbola,* buffalo.

ciclar (zee-KLAR), v., to bike.

ciclin (see-KLEEN), n., m., bicycling. E»S

ciclocross (zee-kloo-KROOS), n., m., dirt biking. E»S

ciclostil (zee-kloo-STEAL), n., m., ciclostyle, e.g., machine that reproduces handwriting and designs. E»S

ciclostilo (zee-klo-STEE-loh), n., m., bicycle style. E»S

ciénaga (ZEE-EH-ne-gah), n., m., swampy region with springs.

cienporcién (sean-por-SEAN), exp., one hundred percent. "He is on a diet cienporcién." S»E

cinc (SYNC), n., m., zinc. {Ch, NY, SW} E»S

cinch (SEENCH), n., m., saddle fastener. Sp. *cincha.*

cincho (ZEEN-tcho), exp., sure thing. "La compra de la casa is un cincho." {Ch}

cizoet (zee-zoh-ETH), n., m., wound. "Tuvo un accidente y sufrió un cizoet." {Ch}

claco (KLAh-ko), nickel, five cents. "Necesito un claco más pa' comprar el newspaper." {Ch}

claimear (klay-MEAR), v., to climb. E»S

claimin (KLAY-mean), n., m., social mobility. "A ella le gusta el claimin." E»S

clanga (KLAN-gah), n., f., gang. "You don't want to mess with mi clanga." {Ch, LA, SW}

clapiar (kla-PEAR), v., to clap, to applaud. {Ch} E»S

claro (KLA-roh), n., m., light spot. The term dates from 1891.

clavar (kla-VAR), v., to steal. "Clávate ese anillo. Steal that ring." {Ch}

clavero (kla-VE-roh), n., m., thief. {Ch}

clavete (kla-VE-te), theft. {Ch}

clearin (CLEA-reen), n., m., commerce system between nations to establish mutual compensation. "Los acuerdos de clearin entre United States y México."

clecha (KLE-tcha), n., f., school. "I went to clecha today." {ELA}

clergiman (kler-gee-MAN), n., m., religious clothes worn by priests, ususally a dark color jacket and black pants. E»S

clica (KLEE-ka), n., m., 1. gang. 2. group of friends. "These friends are part of my clica." {Ch}

clikear (klee-KEAR), v., to click. {CS} E»S Also CLIQUIAR.

cline (KLINE), n., tissue paper. After brand name *Kleenex*. "Los viejos prefieren andar con un pañuelo in their pocket que usar cline." E»S Also KLINEX.

clínica (KLEE-nee-ka), n., f., exercise. "Los niños tuvieron una clínica de soccer." Sp. *clínica* means doctor's clinic. E»S

clip (KLEEP), n., m., paper clip. E»S

clipa (KLEE-pah), n., m., clippers. "Alcánzame la clipa. Hand me the clippers." {G} E»S

clipe (KLEE-peh), n., m., clip. Cyber-Spanglish term meaning to attach. {CS}

clipear (KLEE-pear), v., to type, also to staple. Also CLIPIADORA. {CS} E»S

cliper (KLEE-per), n., m., plane, boat. E»S

clipiadora (klee-pea-DO-ra), n., f., stapler. Sp. *engrapadora*. E»S

clipiar (klee-PEAR), v., to staple. {EC} E»S

cliquiar (klee-KEAR), v., 1. to socialize. 2. to click the mouse of a computer. {CS} E»S Also CLIKEAR.

cloche (KLOO-tche), n., m., clutch, automobile part. "Aplica el cloche when you change gears." E»S

clon (KLON), n., m., 1. clone. 2. clown. E»S

closet (KLOH-set), n., m., closet. E»S

closter (KLOS-ter), n., m., cluster of objects or people. E»S

clube (KLOO-beh), n., m., club. E»S

cobol (ko-BOL), n., m., programming language. {CS}

cocedor (ko-ZE-dor), n., m., oven. {Ch}

cocoa (ko-KO-ah), n., f., the seed of the cocoa tree. The term dates from 1707. S»E

cócono (KO·ko·no), n., m., gigolo. {Ch}

código de acceso (KO·dee·go de a·XEH·soh), n., m., access code. {CS}
S»E

cofibreic (ko·fee·BREAK), n., m., coffee break. E»S

coil (KOIL), n., m., loop. {Ch}

col (KOL), n., m., phone call. E»S

colapsar (ko·lap·ZAR), v., to collapse, to unify. "María colapsó dos doc·
umentos into one." S»E

colectador (ko·lec·ta·DOOR), n., m., tax collector. {Ch} E»S

colectar (ko·lec·TAR), v., to collect. E»S

colegio (ko·LEH·hee·o), n., m., university. "El colegio de Williams."
Sp. *colegio* is school. E»S

coli (KO·lee), n., m., type of dog.

colopeador (ko·lo·pea·DOOR), n., m., person that hangs out.

colopear (ko·lo·PEAR), v., to hang out, to enjoy. {Ch}

Colt (KOLT), n., m., type of revolver.

com (KOM), n., m., comb. E»S

combiar (kom·BEAR), v., to comb. {Ch} E»S

comento (ko·MEN·toh), n., m., commentary. E»S

comfader (com·FAH·der), n., m., close friend, ally. "Juan es mi com·
fader." {Ch}

comfortable (com·for·TAH·ble), adj., comfortable. E»S

comic (KO·meek), n., m., comic strip. E»S

comodidad (ko·mo·DEE·dad), n., f., 1. sellable item. 2. manufacturing
element. Eng. *commodity*. E»S

compa (KOM·pa), n., m., abbreviation of *compadre*.

compac (kom-PAK), n., m., CD, compact disk. E»S

compañero (kom-pa-NEA-roh), n., m., friend, companion.

competitivo (kom-pe-tee-TEE-vo), adj., competitive. E»S

compilar (kom-pee-LAR), n., m., compiler for computer program. {CS} E»S

complección (kom-pleh-XION), n., f., complexion. "Dolores es de complección mestiza." E»S

composet (kom-po-SET), n., m., compost. Recycling remains. E»S

compromiso (kom-pro-MEE-zo), arrangement. Eng. *compromise*. "Don't forget about the compromiso of money you owe." {Ch} E»S

computación (kom-poo-ta-ZION), n., f., computation.

computador (kom-poo-TA-dor), n., m., computer. Also used for words processor. Similar to the Spanish *computadora*. E»S

Con Safos (kon ZA-fos), exp., May God protect us! {SW}

conductor (con-dook-TOR), n., m/f., 1. driver. 2. opera conductor. {Ch}

conectas (ko-NEC-tas), n., f., connection. Used in singular and plural. "You'll have lots of conectas if you graduate from Harvard." E»S

conector (ko-NEK-tor), n., m., connector. {CS}

conferencia (kon-fe-REN-zia), n., f., athletic league. Sp. *conferencia* is lecture. S»E

confidencia (kon-fee-DEN-zia), n., f., confidence. "No pierdas tu confidencia." E»S

confidente (kon-fee-DEN-te), n., m/f., person who is faithful. {Ch} S»E

conflais (kon-FLAYS), n., m., Corn Flakes. "Hey mom, quiero conflais." E»S

confuso (con-FOO-zo), adj., m/f., confused state of mind. "Estás confuso, carnal?" E»S

congal (kon-GAL), n., m., whore house, brothel. {Ch}

congenial (kon-ge-NEHAL), adj., congenial. E»S

consejero (kon-sse-KHE-roo), n., m/f., counsel.

conselar (kon-ze-LAR), v., to counsel. E»S

conselín (kon-she-LEEN), n., m., counseling. E»S

consistente (kon-sees-TEN-te), adv., consistent. S»E

consistir (kon-sees-TEAR), v., to consist. E»S

constipado (kons-tea-PA-do), n., m/f., constipated. {Ch}

constituencia (kon-stee-to-EN-zia), n., f., electorate. Eng. *constituency.* S»E

consulín (kon-se-LEEN), n., m., counseling. E»S

contestar (kon-tes-TAR), v., to contest. "Ella contestó la demanda." Sp. *contestar* is to answer. E»S

contra (KON-tra), n., m/f., person in the opposition, contrarian. Often used in political speech. "Juan was a contra en Nicaragua."

contradanza (kon-tra-DAN-zah), n., f., country dancing.

controler (kon-TROH-ler), n., m/f., comptroller. E»S

controversial (kon-tro-VER-zial), adj., polemical. E»S

contry (KON-tree), n., f., 1. countryside. 2. country music.

convicto (kon-VEEK-to), n., m., prisoner.

coolísimo (koo-LEE-see-mo), exp., cool! E»S Also CULISIMO.

copel (KOH-pel), n., f., couple. E»S

copirit (ko-pee-REET), n., m., copyright. E»S

coque (CO-keh), n., f., Coke. E»S

coquer (KO-ker), n., m., cocker spaniel, a type of dog. E»S

corner (KOR-ner), n., m., corner kick, soccer term. "Viste ese corner?" E»S

corona (ko-ROH-na), n., f., 1. brand of beer. 2. type of long cigar. In Sp. *corona* means crown.

corral (ko-RAL), n., m., enclosure for horses. The term dates from 1582. S»E

corrida (ko-RREE-da), n., f., bull fight. The term dates from 1898. S»E

corriente (ko-RREHEN-te), adj., inferior. "No me importan los productos de corriente. I don't care for inferior products."

corte (KOR-teh), n., f., courtroom. Sp. *tribunal.* Derives from the English word *court.* Example: "Did you go to la corte?" E»S

costumbre (kos-TOOM-bre), n., f., theatrical prop. Eng. *costume.* Sp. *costumbre* means habit. E»S

costumidor (kos-too-mee-DOR), n., m/f., costumer. E»S

costumizar (kos-too-mee-CZAR), v., to customize. E»S

cotaco (ko-TAH-ko), n., m, Kotex, feminine tampon. {Ch}

cote (KOO-teh), n., m., coat. "Ponte el cote porque it's cold." {Ch} E»S

couch (KOUCH), n., m., training coach. E»S

countdown (kaunt-DAUN), 1. to count backwards. 2. preparation for an ultimatum. E»S

coventur (ko-VEN-toor), n., m., joint venture. E»S

coyote (ko-YOH-teh), n., m., 1. prairie wolf. 2. Person in charge of handling undocumented immigrants across the border. From American Indian *coyotl.* The term dates from 1824. S»E

crac (KRAK), n., m/f., athlete with extraordinary abilities. Also CRACK.

crack (KRAK), n., m/f., athlete with extraordinary abilities. Also CRAC.

crakear (kra-KEAR), v., 1. to break. 2. To crack, to hyperventilate. E»S Also CRAQUEAR and CRAQUIAR.

cranque (KRAN-keh), n., m., crank. "No toques that cranque." {Ch} E»S

craque (kra-KE), n., m., cracker, salty cookie. E»S

craquear (kra-KEAR), v., 1. to break. 2. to crack, to hyperventilate. E»S Also CRAKEAR and CRAQUIAR.

craquiar (kra-KIAR), v., 1. to break. 2. to crack, to hyperventilate. E»S Also CRAQUEAR.

crash (KRASH), n., m., stock-market collapse. "No me hables del crash de octubre." E»S

creolo (KREO-loh), n., m., professional pimp. {Arg}

cricket (KREE-ket), n., m., type of sport. "Es el campión del cricket." Also CRIQUET.

criogénico (kreo-GE-nee-ko), n., m/f., cryogenic. E»S

criquet (KREE-ket), n., m., cricket. E»S Also CRICKET.

crismas (KREES-mas), n., f., a Christmas card. "Ella sent me una crismas." Used in place of *Navidades*. E»S Also KRIJMA.

crol (KROLL), n., m., swimming term. Eng. *crawl*. E»S

croquet (kro-KET), n., m., croquet, athletic activity. E»S

cros (CROS), 1. boxing term. 2. cross-country running. E»S

crosear (kro-SEAR), v., to cross paths. E»S

crus (CROOZ), n., m., tourist cruise. E»S

cruseado (kroo-SEA-do), n., m/f., person who is target of cruising. E»S

crusear (kroo-SEAR), v., to cruise. E»S

cuadra (KWA-dra), n., f., small canyon. S»E

cuáquero (KWA-ke-roh), n., m/f., Quaker. "Es que allá viven muchos cuáqueros." E»S

cuarta (KWAR-tah), n., f., short riding whip used by cowboys. Also spelled *quirt*. {SW}

cuay (KWAY), n., m., guy. {Ch}

Cubanism (koo-ba-NIZM), n., m., a word coined by Cubans. Also CUBANISMO.

Cubanismo (koo-ba-NIZ-mo), n., m., a word coined by Cubans. Also CUBANISM.

cucho (KOO-tcho), adj., misbehaved. {Ch}

cuilca (KWEEL-ka), n., f., cover. {Ch}

cuira (KWEE-rah), n., m., quarter, twenty-five cents. {Ch}

cuiza (KUEE-zah), n., f., prostitute. {Ch}

culeado (koo-LEA-do), adj., seriously messed-up. "Don't pay attention to ese huevón culeado."

culear (koo-LEAR), v. 1. to be afraid. From the Spanish *culo,* ass. 2. to calm down and to chill out. From the slang use of *cool*. 3. to copulate.

culi (KOO-lee), n., m., 1. indigenous worker. 2. collie, canine type.

culísimo (koo-LEE-see-mo), exp., cool! E»S Also COOLISIMO.

cuotizar (koo-o-tee-ZAR), v., to set the price of an item. {Ch}

curandera (koo-ran-DE-rah), n., f., healer.

curry (KOO-ree), n., m., Indian spice.

cursor (koo-SOR), n., m., cursor. {CS} E»S

cusi (KOO-zee), n., m., cook. Sp. *cocinero.* {SW}

cutear (koo-TEAR), v., to cut. "Cutea la carne antes de comer." E»S

cuter (KOO-ter), n., m., cutter. E»S

D

daim (DAYM), n., m., dime, ten-cent coin. E»S Also DAIME.

daime (DAY-me), n., m., dime, ten-cent coin. E»S Also DAIM.

dancin (dan-CEEN) n., m., dancing space. E»S

dandi (dan-DEE), n., m., dandy. E»S

dandismo (dan-DEEZ-mo), n., m., Oscar Wilde's ideology. E»S

databanc (data-BANK), n., m., data bank. {CS} E»S

databeis (data-BEYZ), n., f., data base. {CS} E»S

dataglov (data-GLOV), n., m., video game using a glove for sensorial
    activity. {CS} E»S

datarum (da-ta-ROOM), n., m., investment room. E»S

daunsait (dawn-SAIT), n., m., downsize, reduce corporate size. "Su
    compañía es un daunsait en comparación con Microsoft." E»S

dauntaun (DAWN-tawn), n., m., downtown. "Milly trabaja en el daun-
    taun." {CA, NR} E»S

debitar (de·bee·TAR), v., to owe. E»S

débito (DE·bee·to), n., m., 1. debit. 2. debt. "Antonio tiene un débito de $2,000 con Jim." E»S

debut (de·BOOT), n., m., debut. "María hizo su debut." E»S

debutar (de·boo·TAR), v., to debut. E»S

declinar (de·klee·NAR), v., to decline. "Ella declinó la oferta." E»S

decodear (de·ko·DEAR), v., to decode. {CS} E»S

dedlain (ded·LAIN), n., m., deadline. E»S

defolt (de·FOLT), n., m., default. {CS} E»S

deiof (dey·OF), n., m., day off. "Mañana tenemos un deiof." E»S

deit (DEIT), n., f., date. "Juan habló con Rosiat. Tienen una deit for Saturday." E»S

delayarse (de·la·YAR·se), v., to delay, to be delayed. "Francisco se delayó because of el tráfico." E»S

depende (de·PEN·deh), n., m., f., dependent, relative to whom one is financially connected. E»S Also DEPENDIENTE.

dependible (de·pen·DEE·ble), n., m/f., dependable. E»S

dependiente (de·pen·dee·EN·te), n., m/f., dependent, relative to whom one is financially connected. E»S Also DEPENDE.

depo (DEH·po), n., m., station depot. "Ellos tienen su material en el depo." E»S Also DIPO.

deputado (de·poo·TAH·do), n., m., deputy, second·in·command. Sp. *diputado.*

desabilidad (des·ha·bee·lee·DAD), n., f., disability. E»S

desapointar (des·ha·poin·TAR), v., to disappoint. E»S

desapointeo (des·ha·point·THE·o), adj., disappointed. E»S

descharche (des·CHAR·che), n., m., discharge. E»S

descharchear (des-char-CHEAR), v., to discharge. E»S

deservear (de-ser-VEAR), v., to deserve. E»S

deshabilitado (des-ha-bee-lee-TAH-do), n., m/f., disabled. "Juanito fue a Vietnam y quedó deshabilitado." E»S

deshabilitar (des-ha-bee-lee-TAR), v., to disable. E»S

desperado (des-pe-RAH-do), n., m., violent criminal, outlaw. Sp. *desperado*. S»E

desposeído (des-poh-se EE-do), n., m/f., poor. Eng. *dispossessed*. E»S

detergente (de-ter-GEN-te) n., m., detergent. E»S

devotear (de-vo-TEAR), v., to devote. "El devotea su energía al candidato." E»S

dil (DEEL), n., m., deal. "Hicieron un dil." E»S

diler (DEE-ler), n., m., distributor. "Si hay algún problema, speak directamente con el diler." E»S

dipo (DEE-po), n., m., station depot. "Ellos tienen su material en el dipo." E»S Also DEPO.

dirísimo (deer-EE-see-moh), exp., m/f., dearest. E»S

discopob (dis-ko-POB), n., m., disco bar. E»S

disketera (des-ke-TEH-ra), n., f., disk drive. {CS}

dispachar (dees-pa-CHAR), v., to dispatch. E»S

dispensar (dees-pen-SAR), v., to dispense, distribute. E»S

display (dees-PLAEH), n., m., visual ad. E»S

displayador (dees-pla-ya-DOOR), n., m., person in charge of a merchandise display. E»S

displayar (dees-pla-YAR), v., to display.

disposición (dees-poh-zee-ZION), n., f., disposition. "Ella isn't en la disposición de comprar una new house."

disquete (dis·KE·teh), n., m., diskette. {CS} E»S

disturbiar (dees·toor·BEEAR), v., to disturb. E»S

dixi (dee·XEE), n., m., jazz style. E»S

dobliu-dobliu-dobliu (WWW), n., f., World Wide Web. {CS} E»S

dogibag (do·GEE·bag), n., f., doggy bag. E»S

dolby (dol·BEE), n., m., reduction sound devise. E»S

dolema (do·LEH·ma), n., m., dilemma. {NR}

dombo (DOM·boh), exp., unintelligent, stupid. After the Walt Disney character. {CA} Also DUMBO.

domi (DOH·mee), n., m., dummy. E»S

dompin (DOM·peen), n., m., major business sale. E»S

Don (DON), exp., male courtesy name. "Don Calabacito López." The term dates from 1523. S»E

Don Juan (don·JOAN), n., m., seducer of women. "Don't be such a Don Juan." Also MACADAM and MACADAMO.

Doña (DOH·nya), exp., female courtesy name. "She was the legendary Doña Josefa Ortíz de Domínguez." The term dates from 1622. S»E

dona (DO·nah), n., f., doughnut. E»S

dopar (do·PAAR), v., to dope up. {NE}

dopear (DO·peh·ar), v., to dupe up. E»S

dopelgang (do·pel·GANG), n., f., gang individual with split personality. {Ch}

dorman (DOR·man), n., m., doorman. E»S

draft (DRA·fet), n., m., military recruitment. E»S

draftear (DRAF·teh·ar), v., to draft. E»S Also DRAFTIAR.

draftiar (DRAF·teeh·ar), v., to draft. E»S Also DRAFTEAR.

draguear (dra·GUE·ar), v., to dràg. E»S Also DRAGIAR.

draguiar (dra·GUEE·ar), v., to drag. E»S Also DRAGUEAR.

draiv (DRAY·veh), n., m., computer drive. E»S Also spelled *drive*. {CS}

draque (DRA·ke), n., m., alcoholic beverage.

driblar (DREE·bleh·ar), v., to dribble. Used in soccer parlance. E»S

dribler (DREE·bler), n., m/f., dribbler. E»S Also DRIBLAR. E»S

driblin (DREE·bleen), n., m., the act of dribbling. E»S Also DRI·BLAR and DRIBLER.

drilear (DREE·le·ar), v., to drill. E»S

drinkear (dreen·KEHAR), v., to drink. "Drinkéate the juice" {CA} E»S

drivelar (dree·ve·LAR), v., to drive. "Father me drivelo a la escuela." {Ch} E»S

drog (DROOG) n., f., drugstore. "Vamos a la drug a comprar medici·nas." E»S Also DROGUERIA.

droguería (dro·gue·REE·ah), n., f., drugstore. E»S Also DROG.

droguista (dro·GEES·ta), n., m/f., pharmacist. E»S

dropear (DRO·pehar), v., to drop. "Dropié my friend at the parti."

dropin (DOH·peen), adv., the act of dropping an item. E»S

dudiligens (doo·dee·lee·GENZ), n., m., complete financial analysis. "Yo recibí de la compañía un dudiligens. E»S

duityursel (DOO·it·YOOR·selv), exp., "Do it yourself!" {CA} E»S

dumbo (DOM·boh), exp., 1. unintelligent. 2. stupid. After the Walt Disney character. {CA} E»S Also DOMBO.

dumper (DOM·per), n., m., dumper. "Echa la basura en el dumper." E»S

dumpin (DOM·peen), n., m., economics term referring to the act of dumping products. E»S

dutifri (DOO·tee·free), n., m., duty free. "No hay tax en la tienda du·tifri." E»S

E

ebonita (eh‑boh‑NEE‑ta), n., f., ebonite, plastic material. E»S

echamunas (eh‑tcha‑MYU‑naz), exp., "Give me five." {Ch} E»S

ecomarketing (e‑cho‑MAR‑keh‑teeng), adv., business strategy. E»S

ecual (e‑KWAL), adv., equal. E»S

ecualizar (e‑qua‑LEE‑zar), v., to equalize. E»S

editin (EH‑dee‑teen), v., the act of editing a text. {CS} E»S

educacional (e‑doo‑ka‑zio‑NAL), adv., educational. E»S

efeme (e‑FE‑me), n., m., FM radio.

eforte (e‑FOR‑te), n., m., effort. "Manuel hizo un eforte inmeso para ganar." E»S

egotismo (eh‑go‑TEES‑mo), n., m., egotism. E»S

eiderdáun (EY‑der‑dawn), n., f., bed cover. Sp. *edredón.* E»S

ejec (eh‑JEK), n., m., Eng. *eject.* {CS} E»S

electrochoque (e·lek·tro·TCHOK), n., m., electroshock. E»S

elfo (EL·foh), n., m., elf. {CA} E»S

email (EE·mail), n., m., electronic mail. "Te mando el email immediately." Also IMALITO and MANUELITO. {CS} E»S

emailiar (ee·mai·LEAR), v., to e·mail. {CS} E»S

emanei (em·an·EY), n., m., mergers and acquisitions. Financial term. E»S

embarazado (em·ba·ra·ZA·doo), adv., embarrassed. Sp. *embarazada* means pregnant. E»S

embarazar (em·ba·ra·CZAR), v., to embarrass. Sp. *embarazar* means to become pregnant. E»S

embarcadear (em·bar·ka·DEHAR), v., 1. to stand someone up, not to show up to an appointment. 2. to sell out. From Sp. *embarcar* and *dejar embarcado.* "Oye don't embarcadees me tonight because I know Juanqui is going to be there." {CA} Also EMBARCA·DIAR.

embarcadiar (em·bar·ka·DEEAR), v., 1. to stand someone up, not to show up to an appointment. 2. to sell out. Sp. *embarcar* and *dejar embarcado.* "Oye don't embarcadees me tonight because I know Juanqui is going to be there." {CA} Also EMBARCADEAR.

embedeador (em·be·dea·DOR), n., m/f., person who embeds material goods. E»S

embedear (em·be·DEHAR), v., to embed. E»S

employar (em·plo·YAR), v., to employ, to make use. E»S

employer (em·plo·YER), n., m/f., boss. E»S

empresariar (em·pre·sa·RIAR), v., to enter into business. E»S

empresario (em·pre·CZA·reeho), n., m/f., entrepeneur. Originally used for government official. S»E

ému (EH-moo), n., m., emu, type of bird.

enchantamiento (en-chan-ta-MIEN-to), n., m., enchantment. E»S

enchantar (en-CHAN-tar), v., to enchant. E»S.

encriptar (en-CREEP-tear), v., to encrypt {CS} E»S

encripteador (en-creep-tea-DOOR), n., m/f., person who encrypts. E»S

endawdear (en-daw-DEAR), v., to endow. E»S Also ENDAWDIAR.

endawdiar (en-daw-DEAR), v., to endow. E»S Also ENDAWDEAR.

enforzar (en-for-CZAR), v., to enforce. E»S

enjoyar (en-kho-YAR), v., to enjoy. "El enjoya su viaje a London." E»S

enquar (en-koo-AR), v., to queue, to make a line. E»S

enrolador (en-roh-la-DOOR), n., m/f., person who enrolls in a program. E»S

enrolar (en-rroh-LAR), v., to enroll. Also ENROLLAR and ENRO-LLARSE. E»S

enrolarse (en-ro-LAR-zeh), v., 1. to get enrolled. 2. to get involved. E»S

enrollar (en-rroh-YAR), v., to enroll. Also ENROLAR and ENRO-LLARSE. E»S

enrollarse (en-rroh-YAR-se), v., to enroll, in reflexive mode. Also EN-ROLAR and ENROLLAR. E»S

ensamblador (en-sam-blah-DOR) n., m., 1. assembler. 2. computer program designed to translate different systems into a single unified format. {CS} E»S

enter (EN-ter), n., m., enter button. "Cuando terminas aprieta enter. When you're done hit enter." {CS} E»S

envelop (en-ve-LOP), n., m., envelope. E»S

épica (E-pee-ka), epic. Term used to describe a work of art or entertainment. "Es una película épica." E»S

equiti (E·kwee·tee), n., m., patrimony. E»S

equivocaited (e·KEE·vo·KAI·ted), adj., state of being mistaken, wrong. "You're very equivocaited with me si crees que voy a estar yo to' el día encerrada aquí mientras tu andas callejeando." {CA} E»S

esa (ESSA), n., f., female dude. "Orale esa, where you from? Hey esa, what time is it?" The term is used less frequently than its male counterpart ESE. It is also used as pronoun. {ELA}

escalibilidad (es·ka·lee·bee·lee·DAD), n., f., scalability. {CS} E»S

escanear (es·KA·near) v., to scan. {CS} E»S

escáner (es·KAH·ner), n., m., scanner. {CS} E»S

escar (e·KAR), n., m., scar. E»S

escaut (es·KAUT), n., m/f., boy and girl scout. E»S

escolar (es·ko·LAR), n., m/f. scholar. "Ella es una escolar distinguida en Women Studies." Sp. *escolar* means school·related. E»S

escor (ES·kor), n., m/f., 1. escort. Used to refer to paid sexual partner. "Necesito un escor this night." 2. score of a sporting event. "Hey, que fue el final escor?" E»S

escortear (es·kohr·TEAR), v., to escort. E»S

escotch (es·KOTCH), n., m., Scotch whisky. E»S

escout (es·KAUT), n., m., in reference to the Boy Scouts. E»S

escrapas (es·KRAH·pas), n., f., scraps. "Tira las escrapas of the window to the garbage." {NR} E»S

escriptear (es·KREEP·tear), v., to script. Filmmaking term. E»S

escuela alta (es·KWE·la AL·tah), n., f., high school. E»S

escuela elemental (es·KWE·la e·le·MEN·tal), n., f., elementary school. E»S

escuela media (es·KWE·la ME·dia), n., f., middle school. E»S

escultismo (es·kool·TEES·mo), n., m., Boy Scout movement. E»S

escutear (es·KOO·tear), v., to scoot. E»S

escuter (es·KOO·ter), n., m., scooter. E»S

ese (ESSE), n., m., male dude. "Orale ese, where you from? Hey ese, what time is it?" The term is used far more frequently than its female counterpart ESA. It is also used as pronoun. {ELA}

eslogan (es·LO·gan), n., m., slogan. "Did you paint ese eslogan en la pared?" E»S

esmachar (es·mah·TCHAR), v., to smash. E»S

esmog (es·MOG), n., m., atmospheric contamination, polluted fog. "No se ven las montanas por el esmog." E»S

esmokin (es·MOH·keen), n., m., tuxedo. E»S Also ESMOQUIN.

esmoquin (es·MOH·keen), n., m., tuxedo. E»S Also ESMOKIN.

esnack (es·NAK), n., m., snack food, appetizer. E»S

esnifar (es·nee·FAR), v., to sniff. In reference to inhaled drugs. "René esnifa cocaína." E»S

esnob (es·NOB), n., m/f., snob. E»S

esnobismo (es·no·BEES·moh), n., m., snobism, the art of being a snob perfected by Joseph Epstein in the late 20th century. E»S

esnorkel (es·NOR·kel), n., m., tube to breathe underwater. E»S

espanglish (es·PAN·gleesh), n., m., Spanglish. Also CALÓ, PACHUCO, SPANGLISH.

espeleo (es·pe·LEH·o), n., m., the spelling. "No sé el espeleo de la palabra 'espanglish.' " E»S

espeletear (es·pele·TEHAR), v., to spell. E»S Also ESPELIAR.

espeletiado (es·pe·le·TEA·doo), adv., m/f., spelled out. E»S Also ESPELIADO.

espeliado (es·pe·LEA·do), adj., spelled out. E»S Also ESPELETIADO.

espeliar (es·pe·LEAR), v., to spell. E»S Also ESPELETEAR.

espich (es·PEETCH), n., m., speech. {NY, CA, Ch, NR} E»S Also ESPICHE.

espiche (es·PEE·tche), n., m., speech. {NY, CA, Ch, NR} E»S Also ESPICH.

espin (es·PEEN), n., f., physics term. E»S

espiritual (es·pee·ree·TUAL), adv., ingenious. E»S

esplín (es·PLEEN), n., m., spleen. E»S

espolear (es·po·LEAR). v., to spoil. E»S Also ESPOLIAR.

espoliar (es·po·LEEAR), v., to spoil. E»S Also ESPOLEAR.

esponsear (es·pon·ZEAR), v. to sponsor. E»S Also ESPONSORIZAR.

esponsor (es·PON·zor), n., m., sponsor. E»S

esponsorizar (es·pon·zo·ree·ZAR), v., to sponsor. E»S Also ESPON·SEAR.

espor (es·POR), adj., casual wear. {CA} E»S

espray (es·PRAY), n., m., spray. Sp. *aerosol.* {CA} E»S

espresgüei (es·PRESS·wey), n., m., expressway, highway. "Por el espresgüei se llega a Miami right away." E»S

espreyar (es·pre·YAR), v., to spray. {CA} E»S

esprín (es·PREEN), n., m., spring. E»S

esprintar (es·PREEN·tar), v., to sprint. E»S

esprinter (es·PREEN·ter), n., m., sprinter. E»S

esquí (es·KEE), 1. v., to ski. 2. ski gear. E»S

establishear (es·ta·blee·SHEAR), v., to establish. E»S Also ESTAB·LISHIAR.

establishiar (es·ta·blee·SHEAR), v., to establish. E»S Also ESTAB·LISHEAR.

establishment (es-tah-BLEESH-ment), n., m., dominant social group. E»S

estado-de-arte (es-TA-doo-de-ar-TEH), n., m., 1. state-of-the-art. "Esta TV es estado-de-arte." 2. the newest fashion. Sp. *el último grito de la moda.* E»S

estancia (es-TAN-zia), n., f., Spanish American cattle farm. The term dates from 1704. S»E

estándar (es-TAN-dar), n., m., standard. E»S

estanflación (es-tan-FLAH-cion), n., f., stagflation. E»S

estar en el tiviri (es-tar en el TREE-vee-ree), exp., socially busy person. Eng. *activity.* "Esa Raquel no para d'estar en el tiviri!" {CA} E»S Also TIVIRI.

estar en llamada (es-tar-en-lla-MA-dah), exp., to be on call. "This after-noon el doctor está en llamada." E»S

estate (es-TA-teh), n., m., estate, real estate legacy. E»S

estatus (es-TAH-tus), n., m., status. E»S

estavanza (es-ta-VAN-zah), n., m., innocuous verbal addiction. "Ilan es un estavanza. He loves los diccionarios." {CS, Mas}

estereo (es-TEh-reo), n., m., stereo. E»S

estido (es-TEE-do), n., m., financial estimate. E»S Also ESTIMADO.

estilográfica (es-tee-lo-GRA-phee-ka), n., f., fountain pen. Sp. *pluma estilo-gráfica.* From Eng. *stylographic,* which in turn comes from Latin.

estimado (es-tee-MA-do), n., m., financial estimate. Sp. *estimado* means *esteemed.* E»S Also ESTIDO.

estok (es-TOK), n., m., stock. Used in financial markets. E»S

estoqueado (es-to-KEA-do), adj., m/f., to be forced to remain in one place for a period of time. From Eng. *stuck.* "Estuvimos esto-queados en el tráfico for three hours." E»S

estraic (es-TRAIK), n., m., strike. Athletic term. "El bateador dio un estraik." E»S Also ESTRAIK and STRAIK.

estraiquiar (es-tra-kEAR), v., 1. to strike. 2. Baseball term: to strike out.

estrés (es-TRES), n., m., stress. "I was feeling estrés after the tough day." E»S Also STRES.

estrésico (es-TRE-si-ko), adj., related to stress. E»S Also ESTRE-ZADO.

estrezado (es-tre-SA-do), adj., m/f., to be overly nervous, anxious, under enormous pressure. E»S Also ESTRESICO.

estrezar (es-tre-ZAR), v., to be the victim of stress, to stress. E»S

estriptis (es-TRIP-tees), n., m., striptease. "Fuiste al show de estriptis?"

estriquin (es-TREE-keen), n., m., streaking. {CA, NY} E»S

estroc (es-TROK), n., m., stock. E»S

estroncio (es-TRON-sio), n., m., strontium. {CA}

estultificar (es-tool-tee-fee-KAR), v., to stultify. E»S

estultificarse (es-tool-tee-fee-KAR-se), v., to be stultified. E»S

estylo (es-TEE-lo), n., m., style. "Ese cantante tiene estylo." Sp. *estilo.* E»S

exami (e-XA-mee), n., f., camisole. The term dates from 1900. Also CAMMY and CAM.

excusa (ex-KOO-za), n., f., excuse. "I don't have en excusa for not attending to your requests." {Ch}

expertríz (ex-per-TREES), n., f., expertise. "Ella tiene expertriz en lavar ventanas." E»S

explotar (ex-plo-TAR), v., to explode emotionally. "Juan explota de amor for María." E»S

F

fábrica (FA‑bree‑ka), n., f., fabric, texture. In Sp. *fábrica* means factory. E»S

facilidad (fa‑zee‑lee‑DAD), n., f., 1. compound, building facility. 2. rest room, toilet. In Sp. *facilidad* means easiness. E»S

factor (FAK‑tor), n., m/f., 1. factory worker. 2. piece of information. E»S

factoría (fak‑to‑REE‑a), n., f., factory. E»S

factorín (fak‑to‑REEN), adv., manufacturing endeavor. "Mi compañía contrató una empresa de factorin para que cobrara las facturas no pagadas." E»S

factual (fak‑TUAL), adv., factual. "Esta historia es factual." E»S

facultad (fa‑KOOL‑tad), n., f., academic faculty. In Sp. *facultad* means a departmental building where the faculty is housed, as in Facultad de Filosofía y Letras. E»S

fagueta (fa‑GE‑tah), n., f., homosexual. Eng. *faggot.* E»S

fail (FAIL), n., m., file. "Abrí el fail en la computadora." E»S

faite (PHAI·te), n., f., fight. {CA} E»S

faiter (fai·TER), n., m., battle plane. E»S

fakería (fa·ke·REE·a), n., f., imitation, cheap reproduction. E»S Also
  FAQUERÍA

falacia (pha·la·ZIA), n., f., fallacy. E»S

falso (FAL·so), n., m., undercover agent. "El estuvo trabajando de falso
  en Irak for seven years." {Ch}

famalirun (fa·ma·lee·ROON), n., m., family room. {CA} E»S

fan (FAN), n., m/f., aficionado. E»S

fandango (fan·DAN·go), n., m., 1. wild dance. 2. fight. Originally a
  particular type of Spanish dance. The term dates from the 18th
  century. S»E

fanzine (fan·ZEEN), n., f., magazine targeted to aficionados of a pop
  star or specific sport. E»S

faquería (fa·ke·REE·a), n., f., imitation, cheap reproduction. E»S Also
  FAKERIA.

fasfiter (fas·PHEE·ter), n., m., plumber. {Chi} E»S

fasfú (fas·FOO), n., f., fast food. E»S Also FASTFOOD.

fashonable (fa·sho·NA·bleh), adj., fashionable. E»S Also FASIO·
  NABLE.

fasionable (fa·zio·NA·bleh), adj., fashionable. E»S Also FASHO·
  NABLE.

fastfood (fast·FOOD), n., f., fast food. E»S Also FASFU.

fastidioso (fas·tee·DEEO·zo), adj., m/f., fastidious, detailed. Sp. *fastidioso* means obnoxious. E»S

faul (FAWL), adj., 1. repugnant. 2. out of bounderies. Athletic term. "El bateador dio un faul." E»S Also FOUL.

faundear (fawn·DEHAR), v., to find. E»S Also FAUNDIAR.

faundiar (fawn·DIHAR), v., to find. E»S Also FAUNDIAR.

featura (fe·a·TOO·rah), n., f., feature. "La featura de esa casa es el gran tamaño." E»S

feelin (FEE·leen), n., m., 1. emotion. "Siento un feelin por Alison." 2. plastic surgery involving the injection of collagen in implants. "Ella se hizo un feelin en los senos." E»S Also FÍLIN.

felo (PHE·lo), n., m., fellow. "César es un buen felo." E»S

felonía (fe·lo·KNEE·a), n., m., felony. E»S

feria (FE·ria), n., f., money. "I need some feria for tonight." {Mex, SW}

ferplei (fer·PLEY), exp., fair play. E»S

ferry (FE·ree), n., m., ferry·boat. "Vamos a tomar el ferry to cross the strait." E»S

feuda (FEU·da), n., f., dispute. "Antonio e Hilda tienen una feuda." {Ch} E»S

ficción (fee·XION), n., f., imaginative literature. "Borges escribió mucha ficción." E»S

fidbac (feed·BAK), n., m., feedback. E»S

fiesta (fee·ES·ta), n., f., party.

fiftififti (feef·TEE, feef·TEE), exp., unsure, iffy. {CA} E»S

filear (fee·LEAR), v., to file. {CS} E»S

filero (fee·LE·roh), n., m., switchblade. "There comes the man with the filero that dissed me." {CA}

filibuster (phee·lee·BOOS·ter), n., m., filibuster, orchestrated effort to stop a procedure. Sp. *filibustero,* a type of 17th·century pirate in the Antilles. Applied to all buccaneers. The term is associated with the U.S. Senate and is commonly used to describe an attempt to obstruct or delay legislation. E»S

fílin (FEE-leen), n., m., 1. emotion. "Siento un fílin por Alison." 2. plastic surgery involving the injection of collagen in implants. "Ella se hizo un fílin en los senos." E»S Also FEELIN.

filme (FEEL-me), n., m., movie. "Yesterday vimos el nuevo filme de Martin Scorsese." E»S

final-four (fai-nal FOR), exp., four winning athletic teams; sports term. E»S

finca (FIN-ka), n., f., real-estate property. Used also to refer to apartments and small urban houses. "Tenemos una finca en Amesterdam Avenue." S»E

finetunin (fain-TOO-neen), adv., fine-tuning. E»S

finge (FEEN-ge), n., m., 1. offensive gesture. To give the finger. 2. extendable tube in airport terminal. {Ch} E»S

finta (FEEN-ta), n., f, faint. "Ella tuvo una finta." Sp. *finta* means trick. E»S

fintar (feen-TAR), v., to faint. E»S Also FINTEAR and FINTIAR.

fintear (FEEN-tee-ar), v., to faint. E»S Also FINTAR and FINTIAR.

fintiar (FEEN-tear) v., to faint. E»S Also FINTAR and FINTEAR.

firme (FEER-me), exp., alright. "That dance was firme." {Ch}

fisión (fee-ZION), n., m., division of an atom at the nucleus, fission.

fitnes (feet-NES), n., f., fitness. E»S

fixin (FEE-xeen), n., m., financial arrangement. Economic term. E»S

flamear (fla-MEAR), v., to set aflame. E»S

flas (FLAS), n., m., bright light, flash. E»S Also FLASH.

flash (FLASH), n., m., bright light, flash. E»S Also FLAS.

flashbac (flash-BAK), n., m., restrospective look. "El filme está lleno de flashbacs." E»S Also FLASHBACK.

flashback (flash-BAK), n., m., restrospective look. "El filme está lleno de flashbacks." E»S Also FLASHBAC.

flaslai (flaz-LAY), n., f., flashlight. E»S

fletear (fle-TEAR), v., to walk the streets. {CA} E»S

fletera (fle-TE-ra), n., female flirt. "She's tremenda fletera in the streets of Havana." E»S

fliado (flee-A-do), adj., m/f., on loan. {Ch, SW}

fliliar (flee-LEEAR), v., to lend. "Ella flilea dinero."

flica (FLEE-ka), n., f., movie theater, drive-in. Eng. *movie flick.* "Let's go to a flica tonight."

flipar (flee-PAR), v., to be extremely surprised, shocked. "Me flipé when I was fired." E»S Also FLIPEAR.

flipear (flee-PEAR), v., to flip. E»S Also FLIPAR.

flipearse (flee-PEAR-se), v., to flip out. "Creí ver un ghost y me flipié." E»S

fliper (FLEE-per), n., m., video game in the form of an arcade. E»S

flirt (FLEERT), n., m., flirt. E»S

flirteo (fleer-TE-oh), n., m., the act of flirting. E»S

flodear (flo-DEHAR), v., to irrigate, to flood. E»S Also FLODIAR and FLUDEAR.

flodiar (flo-DEHAR), v., to irrigate, to flood. E»S Also FLODEAR and FLUDEAR.

flonkear (flon-KEHAR), v., to flunk an examination. "Flonké el exam." E»S Also FLONQUEAR and FLOQUIAR.

flonquear (flon-KEHAR), v., to flunk an examination. "Flonqué the exam." E»S Also FLONKEAR and FLONQUIAR.

flonquiar (flon-KEHAR), v., to flunk an examination. "Flonqué the exam." E»S Also FLONKEAR and FLONQUEAR.

flopi (FLO-pee), n., m., floppy disk. {CS} E»S

floshear (flo-SHEHAR), v. to flush the toilet, from the English. "Floshea el toilet." E»S

flosheo (flo-SHE-o), n., m., act of toilet flushing. E»S

flowback (flow-BAK), n., m., revenue. Economic term. E»S

fludeador (flu-dea-DOOR), n., m., irrigator, person who floods a site. E»S

fludear (flu-DEAR), v., to irrigate, to flood. E»S Also FLODEAR and FLODIAR.

fob (FOB), n., m., mercantile clause making the buyer pay the price of transportation of shipment but not insurance. Eng. *freight-on-board.* E»S

folclor (fol-KLOR), n., m., folklore. From Eng. and German.

folclórico (fol-KLO-ri-ko), adj., m/f., relative to folklore.

folclorista (fol-klo-REES-ta), n., m., 1. person devoted to folklore. 2. person whose manners become folkloric.

foldear (fol-DEAR), v., to create a folder. {CS} E»S

folder (FOL-der), n., m., folder. "Te enviaré un folder with the document por e-mail." {CS} E»S

foni (FO-nee), exp., funny. "Esa situación está foni." E»S

footin (FOO-teen) v., jogging. E»S Also FUTIN, FUTINEAR, FUTINIAR and JOGIN.

forceador (for-sea-DOR), n., m/f., person forced into a situation. E»S

forcear (for-SEAR), v., to force. "El forceó su presencia entre los amigos." E»S

forcin (FOR-seen), v., to apply pressure. E»S

foreign market (FOREIN-mar-ke), n., m., foreign market. E»S Also FOREINMARQUE.

foreinmarque (FOREIN-mar-ke), n., m., foreign market. E»S Also FOREIGN MARKET.

forguadeador (for-war-dea-DOR), n., m/f., person in charge of forwarding documents. {CS} E»S Also FORWARDEADOR.

forguadear (for-WAR-dear), v., 1. to forward a document. {CS} 2. to pass the ball forward. Athlentic term. E»S Also FORWADEAR.

forguar (FOR-war), n., m. 1. forward button. 2. offensive athletics position. 3. Future-looking contract. E»S Also FORWARD.

forguardeo (for-war-DE-oh), adv., 1. the act of forwarding a document. {CS} 2. The act of forwarding the ball in an athletic activity. E»S Also FORWARDEO.

forguetear (for-gue-TEHAR), v., to forget. E»S Also FORGUETIAR.

forguetiar (for-gue-TEHAR), v., to forget. E»S Also FORGUETEAR.

forleiri (for-LE-ree), n., f., forelady. "Since she's the boss's favorite, se cree la forleiri." {NY} E»S

forma (FOR-ma), n., f., legal document. Eng. *form.* E»S

fornido (for-NEE-do), adj., m., furnished. E»S Also FURNIDO.

fornitura (foor-nee-TOO-rah), n., f., furniture. E»S Also FURNITURA.

fortran (for-TRAN), n., m., programming language used primarily to resolve scientific and technical problems. Acronym of *formula translator.* {CS} E»S

forwadear (for-WA-dear), v., 1. to forward a document. {CS} 2. to pass the ball forward. Athletic term. E»S Also FORGUADEAR.

forward (FOR-ward), n., m. 1. forward button. 2. offensive athletics position. 3. future-looking contract. E»S Also FORGUAR.

forwardeador (for-war-dea-DOOR), n., m/f., person in charge of forwarding documents. {CS} E»S Also FORGUARDEADOR.

forwardeo (for-war-DE-oh), adv., 1. act of forwarding a document. {CS} 2. act of forwarding the ball. Athletic term. E»S Also FORGUARDEO.

fotostática (fo-tos-TA-tee-ka), n., f., Xerox. E»S

foul (FAWL), adj, 1. repugnant. 2. out of bounds. Athletic term. "El bateador dio un faul." E»S Also FAUL.

foundear (fawn-DEAR), v., to find. E»S Also FOUNDIAR.

foundiar (fawn-DEAR), v., to find. E»S Also FOUNDEAR.

fraseología (fra-seo-lo-GEE-a), n., f., phraseology. E»S

freak (FREEK), n., m/f., unstable individual. E»S Also FRIC.

free (FREE), exp., gratis. The term dates from the mid 18th century. E»S Also FRI.

freelance (free-LANZ), n., m/f., work-for-hire employee. E»S Also FRILANS.

freezar (free-CZAR), v., to freeze. E»S Also FRIZAR and FRIZEAR.

freezer (FREE-zer), n., m., refrigerator. E»S Also FRIZER, FRIGIDER and FRIGIDAIRE.

freno (FRE-noh), n., m., bridle for a horse. S»E

fresa (FRE-sah), adj., formal, strict. {Ch, M}

fresca (FRES-ka), n., f., soda drink. Also SODA.

fresco (FRES-ko), adj., bad-mouthed.

fresh (FRESH), exp., in-style, of good taste. E»S

fri (FREE), exp., gratis. The term dates from the mid 18ᵗʰ century. E»S Also FREE.

fric (FREEK), n., m/f., unstable individual. E»S Also FREAK.

frigidaire (FREE-hee-dayr), n., m., refrigidator. E»S Also FREEZER, FRIZER and FRIGIDER.

frigider (FREE-hee-der), n., m., refrigerator. (Chi) E»S Also FREEZER, FRIZER and FRIGIDAIRE.

frijoles (free-KHO-les), n., m., beans. "The Sánchez family diet is arroz with frijoles." S»E

frilancear (free-lan-ZEAR), v., to freelance. E»S Also FRILANZAR.

frilans (free-LANZ), n., m/f., work-for-hire employee. E»S Also FREELANCE.

frilanzar (free-lan-ZAR), v., to freelance. E»S Also FRILANCEAR.

friqui (FRE-kee), n., m., free kick. Soccer term. E»S

fritado (free-TAH-do), adj., m., fried food. "A Marisol le gustan los fritados." S»E Also FRITANGA.

fritanga (free-TAN-ga), n., f., fried food. "A Marisol le gustan las fritangas." S»E Also FRITADO.

frizado (free-ZA-do), adj., m/f., paralyzed, shocked. "Se quedó frizado with the news!" E»S Also FREZEADO and FRIZIADO.

frizar (free-CZAR), v., to freeze. E»S Also FREEZAR and FRIZEAR.

frizeado (free-ZEA-do), adj., m/f., 1. paralyzed. 2. shocked. "Se quedó frizeado with the news!" E»S Also FREZEADO and FRIZIADO.

frizear (free-CZEAR), n., m., to freeze. E»S Also FREEZAR and FRIZAR.

frizer (FREE-zer), n., m., refrigerator. E»S Also FREEZER, FRIGIDER and FRIGIDAIRE.

friziado (free-ZEE-A-do), adj., 1. paralyzed. 2. shocked. "Se quedó fri-ziado with the news!" E»S Also FREZEADO and FRIZADO.

ful (FOOL), n., m., 1. foul shot. Athletic term. 2. gas. Eng. *fuel.* E»S

fuldisclos (fool-dis-KLOS), exp., full disclosure. Legal term. E»S

fulear (foo-LEAR), v., to be in a hurry. E»S

full (FOOL), n., m., winning hand. Poker term. E»S

full contact (fool KON-tac), n., m., rough athletic encounter. S»E

full time (FOOL-taym), exp., full-time employee. E»S Also FUL-TAIM.

fultaim (FOOL-taym), exp., full-time employee. E»S Also FULL TIME.

fun (FON) adj., filled with joy. "La pasé fun en la fiesta."

funbor (FON-bor), n., m., fun board, a type of windsurf done over a board.

fund (FOND), n., m., management fund. "Vamos a comprar el fund" or "She is una administradora de funds."

fundear (fun-DEAR), v., to fund financially. E»S

fundin (FUN-dean), n., m., support money. E»S

fundlear (fun-DLEAR), v., to excite sexually. E»S

fundlin (FUN-dlean), exp., sexually excited. E»S

funk (FONK), n., f., popular modern music. E»S Also FUNKY.

funky (FON-kee), exp., popular modern music. E»S Also FUNK.

furnido (foor-NEE-do), adj., m., furnished. E»S Also FORNIDO.

furnitura (foor-nee-TOO-rah), n., f., furniture. E»S Also FORNI-TURA.

fuselaje (foo-se-LA-hke), n., m., fuselage. E»S

fusión (FOO-shion), n., f., music interjecting jazz and rock. E»S

fútbol (FOOT-bol), n., m., soccer. {SW} S»E

futbolista (foot-bo-LEES-ta), n., m/f., soccer athlete. {SW} S»E

fútin (FOO-teen) v., jogging. E»S Also FOOTIN, FUTINEAR, FU-
TINIAR and JOGIN.

futinear (foo-tee-NEAR), v., jogging. E»S Also FOOTIN, FUTIN,
FUTINIAR and JOGIN.

futiniar (foo-tee-NEE-ar), v., jogging. E»S Also FOOTIN, FUTIN-
EAR, FUTIN and JOGIN.

G

gabacho (ga·BAH·tcho), n., m/f., white person. {Ch}

gabinete (ga·bee·NE·te), n., m., 1. political committee. 2. kitchen cabi-
    net. Interchangeable with *pantri*. "I keep the pots in the gabi-
    nete de arriba." E»S

gacho (GAH·tcho), exp., messed·up, in poor shape. "¡Qué gacho, your
    lady left you!" {Ch}

gag (GAG), n., m., 1. performance. 2 act. 3. situation. "Jaime es bueno
    pa' los gags." E»S

gagman (GAG·man), n., m., comedian. E»S

gaipe (GAI·peh), n., m., burlap. {Chi}

gait (GAIT), n., f., gate. E»S

galei (ga·LEY), n., f., bar on an airplane or train.

galón (ga·LON), n., m., gallon. Measurement unit.

game (GAYM), exp., 1. athletic talent. "El tiene un buen game." 2. conclusion of a an athletic activity. "El marcador es 18–26 y allí quedó el game." E»S

ganga (GAN-ga), n., f., 1. gang. "Are you part of any urban ganga?" 2. sale. "Hay gangas en la T.J. Maxx store." E»S

gangster (gan-STER), n., m., mafioso. E»S

gap (GAP), n., m., division. "Hay un gap entre el father y la daughter." E»S

gar(r)otte (ga-ROH-teh), n., m., 1. police stick. 2. nasty person. Sp. *garrote* and Eng. *garrotte*. The term dates from 1850. S»E

garaje (ga-RA-geh), n., m., garage. From Fr. *garage.* E»S

gasetear (ga-se-TEAR), v., to put gas into an automobile. E»S

gasetería (ga-se-te-REE-ah), n., f., gas station. E»S

gasfiter (GAS-fee-ter), n., m., gas attendant. E»S

gasoleo (ga-so-LEH-o), n., m., gasoline. E»S Also GASOLIN.

gasolín (ga-so-LEEN), n., f., gasoline. E»S Also GASOLEO.

gasquet (gas-KET), n., f., gasket. E»S

gastar el tiempo (gas-TAR el TEE-EM-poh), exp., to waste time. E»S

gaucho (GAU-tcho), n., m., provincial person, cowboy. The term dates from 1824. S»E

gay (GEY), n., m/f., homosexual person. "Reinaldo Arenas era gay." E»S Also GEY.

gazpacho (gas-PA-tcho), n., m., cold vegetable soup. The term dates from 1845. S»E

gentelman (gen-tel-MAN), n., m., respectful male. Sp. *caballero.* "El Señor Sobejano es todo un gentelman." E»S Also GENTLEMAN.

gentleman (gen·tle·MAN), n., m., respectful male. Sp. *caballero.* "El Señor Sobejano es todo un gentleman." E»S Also GEN·TELMAN.

gey (GEY), n., m/f., homosexual person. "Reinaldo Arenas era gey." E»S Also GAY.

gibón (gee·BON), n., m., gibbon. E»S

gilet (gee·LET), n., m., razor blade, after brand Gillette. E»S

gimjaz (GEEM·jaz), n., m., type of gymnastics practiced with jazz music. E»S

gimkana (gym·KA·nah), n., m., competition in which the participants overcome obstacles and difficulties, especially involving vehicles. {Ch}

gin (GEEN), n., m., alcoholic drink. E»S

gingerale (geen·ger·ALE). n., m., soft soda drink, made of ginger. E»S Also GINGEREIL.

gingereil (geen·ger·ALE), n., m., soft soda drink, made of ginger. E»S Also GINGERALE.

gintonic (geen·TO·nik), n., m., mixed alcoholic drink, made of gin and tonic water.

girl (GIRL), n., f., woman acting as a ballerina in a variety show. "El show included varias girls en la sección de baile." E»S

girlscaut (girl·SCAUT), n., f., girl scout, outdoor girl. E»S

glam (GLAM), exp., panache of a music group. "El aire glam de este grupo me resulta desagradable." {ELA,T}

glamour (gla·MOOR), n., m., style. Fr. *glamour.* Also GLAMUR.

glamur (gla·MOOR), n., m., style. Fr. *glamour.* Also GLAMOUR.

globear (glo·BEAR), v., around·the·globe travel. Also GLOBETRO·TEAR.

globero (glo-BE-roh), n., m/f., person who travels around the globe. Sp. *globero* means *baloon maker*. Also GLOBTROTERO.

globtroter (glob-tro-TEAR), v. to travel around the world. "Juan glob-trotea en su nuevo trabajo." E»S Also GLOBEAR.

globtrotero (glob-tro-TEH-roh), n., m/f., person who travels around the globe. E»S Also GLOBERO.

gocaso (go-KA-so), n., m., to take a room or a flat and become a prosti-tute. "She only married him for the money, moved in and be-came a gocaso." Also CASO.

gofres (GO-fres), n., m., waffles. {S}

gogo (GO-go), n., f., funky girl. "Ella es una chica gogo." E»S

gol (GOL), n., m., 1. goal, objective. 2. goal kick. Soccer term. E»S

gol-average (GOL a-ve-rag), n., m., goal average. Athletic term. E»S

golf (GOLF), n., m., golf. "Voy a jugar golf con el Presidente Clinton esta tarde." E»S

gom (GOM), n., m., bubble gum. "Tienes más gom to chew?" E»S

gong (GONG), n., m., gong. Orig. from Malaysian. E»S Also GONGO.

gongo (GON-go), n., m., gong. From Malaysian. Also GONG.

gor (GOR), exp., gory. "Esta película de horror es gor." E»S Also GORE.

gore (GO-ree), exp., gory. "Esta película de horror es gore." E»S Also GOR.

goretex (GO-re-tex), n., m., Scotch tape. After brand name. "Ella usa goretex en sus maquetas." E»S

gospel (GOS-pel), n., m., type of religious music. "Ella canta el gospel en la church." E»S

grafeo (gra‑PHE‑oh), n., m., graffiti design. {SW, NY, T} E»S Also
   GRAFITO.

grafiar (gra‑FEAR), v., to graph, to handwrite. E»S

grafitear (gra‑fee‑TEHAR), v., to make graffiti. E»S

grafito (gra‑FEE‑to), n., m., graffiti design. {SW, NY, T} E»S Also
   GRAFEO.

gramy (GRA‑mee), n., m., as in the music award. "Rubén Blades fue
   nominado pa' un gramy." E»S

grande (GRAN‑de), adj., 1. large‑size coffee. 2. large‑size penis. S»E

grandee (gran‑DEE), n., f., big. S»E

gratificación (gra‑tee‑fee‑ka‑ZION), n., f., gratification. "Lo impor‑
   tante en la vida no es únicamente la gratificación." E»S

greaser (GREE‑ser), n., m., person of mixed race, frequently of Mexican
   and Indian origin. {Ch, SW} E»S

green (GREEN), n., m., 1. garden. 2. yard. 3. golf course. E»S Also
   GRIN.

greenfreeze (GREEN‑freez), n., m., ecologically‑sound freezing ma‑
   chine used by corporations. E»S

greifrut (GREY‑frut), n., f., grapefruit. E»S Also GREITFRUTA.

greivi (GREY‑vee), n., m., gravy. "La carne viene con un greivi sabroso."
   E»S Also GREVE.

greve (GRE‑ve), n., m., gravy. "La carne viene con un greve sabroso."
   E»S Also GREIVI.

grifa (GREE‑pha), n., f., marihuana. {SW} Also RIFER.

gril (GREEL), n., m., grill. E»S

grin (GREEN), n., m., 1. garden. 2. yard. 3. golf course. E»S Also
   GREEN.

grincar (GREEN-kar), n., f., Green Card. Document given to newcomers to the United States who receive legal status by the Immigration and Naturalization Service [INS]. E»S

gringo (GREEN-go), n., m., f., 1. white American. 2. foreigner. The origin might refer to gibberish. Or else, to green-go, as in green-money goes fast. Another etymological version is based in the green-color uniform of American soldiers abroad. The term dates from 1846. S»E

gringófobo (green-GO-pho-bo), n., m/f., a person who dislikes Gringos. "She refuses to visit the United States because she's a gringófoba." {Ch}

gringólatra (green-GO-la-tra), n., m/f., a person who adores Gringo culture. {Ch}

Gringoñol (green-go-NYOL), n., m., mestizo language, part English, part Spanish, used predominantly in the United States since WWII. Also CASTEYANQUI, INGLANOL and SPANGLISH.

grocear (gro-SEAR), v., to acquire groceries. E»S Also GROCIAR.

grocería (gro-se-REE-ah), n., f., grocery. "Ellos compraron las grocerías." In Sp. *grocería* means rudeness, profanity. E»S

grociar (gro-SEE-ar), v., to acquire groceries. E»S Also GROCEAR.

grog (GROG), n., m., rum mixed with water. Alcoholic beverage. E»S

grogui (GRO-kee), adj., groggy. E»S

grompi (GROM-pee), adj., troublesome, problematic. "El hermano de Alberto es grompi." E»S

grupi (GRU-pee), n., m/f., member of a tight group of left-wing comrades. "Marina es una grupi." E»S

grupware (GRUP-wer), n., m., informational program facilitating group work. {CS} E»S

grumpi (GRUM-pee), n., m/f., mature individual in a high economic position who competes with younger professionals. "En la oficina el Señor Jiménez es un grumpi. No deja que los jóvenes avancen en la escala laboral." {CA} E»S

grunge (GRUN-ge), exp., in fashion. "Esa falda está grunge."

guacamole (wa-ka-MO-leh), n., m., Mexican appetizer salsa made with avocado. S»E

guacha (GUA-tcha), exp., watch out. E»S Also WACHA and WA-CHALE.

guachear (WHA-tchear), v., 1. to observe. 2. to watch out. E»S Also GUACHIAR and WATCHEAR.

guacher (WA-tcher), v., observer. E»S

guachiar (WHA-tchear), v., 1. to observe. 2. to watch-out. E»S Also GUACHEAR and WACHEAR.

guachimán (wa-tchee-MAN), n., m., guard, night watchman. E»S Also HUACHIMAN and WACHIMAN.

guafe (WA-fe), n., m., wharf. E»S

guafle (WA-fle), n., m., waffle. E»S Also WAFLE.

guaflear (WA-flear), v., to eat waffles. E»S Also WAFLEAR.

guaipe (WAI-peh), n., m., windshield wiper. E»S Also WAIPE.

guaipear (wai-PEAR), v., to clean the windshields. E»S

guanguei (WAN-wey), n., m., one way. E»S

guardavidas (war-da-VEE-das), n., m/f., swimming-pool life guard. E»S

guardo (WAR-doh), n., m., 1. officer commanding a guard-ship, 2. a petty officer in charge of a penitentiary. The term dates from the late 18th and early 19th centuries.

guasá (wa-SA), exp., 1. what's up? 2. what's happening? "Wasá, hombre, how are you?" 3. joke. Also WASA. {Ch, SW} E»S

guasup (waz-UP), exp., What's up? E»S

güat (GUAT), exp., What? E»S

guava (GUA-vah), n., f., type of fruit. The term dates from 1555. S»E

guaya (GUA-yah), n., f., wire. E»S

güe (GUE), exp., friend. {Ch, Mex} Also BROTHER, CARNAL, GÜEY, VATO, and CARNAL.

güeja (WEE-hka), n., f., head. {SW}

guelda (GUAL-dah), n., f., welder. E»S

guerrilla (gue-REE-ya), n., m., small revolutionary unit. The term dates from 1809 and was popularized in the Spanish Civil War. Also spelled *guerilla*. S»E

güey (gu-EY), n., m., friend. "Orale, güey, join me to the party." Also BATO, BRO, BRODER, BROTHER, CARNAL, GÜE, and VATO.

gufeado (gu-FEA-o), adj., m/f., 1. fashionable, excellent. 2. an object of one's liking. "Esa RANFLA está gufeada." E»S

gufear (gu-FEAR), v., to joke, to kid. From Eng. *goof.* E»S

Güijoken (wee-KHO-ken), n., Weehawken, New Jersey. "Casi todos los cubanos si no vivían en Union City vivían en Güijoken." {CA} E»S

güila (WEE-la), n., f., 1. kite. 2. loose woman. 3. prostitute. {Ch}

güila galetera (WEE-la ga-le-THE-ra), exp., loose woman from the galeta section of town. {Ch}

güinchil (WIN-tchil), n., m., wind-shield. E»S Also WINCHIL.

guinea (wee-NE-ah), n., f., from English guinea, which is an old English gold coin.

guipear (wee-PEAR), v., to wipe. E»S Also GUIPIAR.

guipiar (wee-PEE-ar), v., to wipe. E»S Also GUIPEAR.

guisa (WEE-sa), n., f., good-looking woman. "Mercedes, in spite of her age, es una guisa." {Ch}

güisqui (WEES-kee), n., m., whisky. E»S Also WISKI.

gusa (WOO-sah), exp., 1. what's up? 2. what's happening? "Gusa, en qué andas?" 3. v., f., joke. "Maribel hizo una gusa that made me laugh." {Ch, SW} E»S Also WASÁ.

gusano (goo-SA-no), n., m/f., anti-Castro Cuban. The reference is un-empathetically targeted toward Cuban individuals in Miami. S»E

H

hacienda (ha-SIEN-dah), n., f., ranch estate. The term dates from 1760. S»E

haibol (KHAI-bol), n., m., high-ball. Mixed alcoholic beverage. E»S Also JAIBO and HIGH-BALL.

haifenado (hai-fe-NAH-do), adv., m., hyphenated. E»S

hakear (HA-kear), v., to hack. {CS} E»S Also JAQUEAR and JAKEAR.

haker (HA-ker), n., m/f., computer hacker, Internet pirate. {CS} E»S Also HAQUER and JAKER.

hafcort (haf-CORT), n., m., a sport similar to tennis practiced in a smaller court. E»S

halftaim (HALF-taim), n., m., part time. E»S

hamaca (ha-MA-ka), n., f., hammock. S»E

hamburger (HAM-bur-ger), n., f., hamburger. E»S Also HAMBUR-GUESA.

hamburguesa (ham-bur-GE-sa), n., f., hamburger. E»S Also HAM-BURGER.

hamburguesería (ham-bur-ge-se-REE-ah), n., f., hamburger joint. E»S

handicap (han-DEE-kap), n., m., obstacle, impediment. "Esa sección del parking es para la gente handicap." {S} E»S

handicapear (han-dee-ka-PEAR), v., to place an obstacle. {S} E»S

handimán (han-dee-MAN), n., m., handy man. E»S Also JANDIMAN

handlin (HAN-dlin), n., m., performed services. "El artículo cuesta $18.99, más el handlin." E»S

hangar (han-GAR), v., to hang out. E»S Also HANGUIAR and HANGEAR.

hangear (han-GEHAR), v., 1. to hang out. 2. to loiter. "Please Papi, sólo vamos a hangear en el beisman de Flaco . . ." E»S Also HANGAR, HANGUIAR, JANGUEAR and JANGUIAR.

hanguiar (han-GEE-ar), v., 1. to hang out. 2. to loiter. "Please Papi, sólo vamos a hanguiar en el beisman de Flaco . . ." E»S Also HANGAR, HANGEAR, JANGUEAR and JANGUIAR.

hapenin (ha-peh-NEEN), n., m., event. E»S

hapiend (ha-pee-END), n., m., happy end. E»S

hapyauer (ha-pee-ah-UER), n., f., happy hour. E»S

haquer (HA-ker), n., m/f., computer hacker, Internet pirate. {CS} E»S Also HAKER and JAKER.

hardcor (HARD-kor), n., m., porno film. E»S

hardguer (HARD-wer), n., m., hardware. E»S Also HARDWER.

hardwer (HARD-wer), n., m., hardware. E»S Also HARDGUER.

hashear (ha-SHEHAR), v., to hash. E»S

hatajo (ha-TAH-jo), n., m., 1. pack of animals. 2. subterfuge. S»E

hauntear (haun-TEHAR), v., to haunt. E»S

havear (kha-VEAR), v., to have. "Ella havea un automóvil green." Sp. *tener* means to have. E»S Also HAVIAR, JAVEAR and JA-VIAR.

haviar (kha-VEAR), v., to have. "Ella havia un automóvil green." Sp. *tener* means to have. E»S Also HAVEAR, JAVEAR and JA-VIAR.

hedge (HEDG), n., f., currency exchange. "Necesito un hedge de pesos a dólares." E»S

hedhonter (HED-jon-ter), n., m., head hunter, talent seeker. E»S

helmet (HEL-met), n., m., helmet. Sp. *casucha, choza.* E»S

hesitación (he-see-ta-TION), n., f., hesitation. E»S

hesitar (he-see-TAR), v., hesitate. "Hesitié antes de responder." E»S

hevy (HE-vee), n., m/f., 1. heavy. "La tarea escolar está heavy." 2. aggressive. "Está música rock es heavy." E»S

hidalgo (ee-DAL-go), n., m. Spanish gentleman. The term dates from 1584. "Don Quixote was an hidalgo from La Mancha." S»E

hidrante (he-DRAN-te), n., m., hydrant. "Los bomberos usan el hidrante para obtener agua en la villita." E»S

hidrobobear (he-dro-bo-BEHAR), v., to float through rough waters. E»S

hidrofoil (he-dro-FOIL), n., m., flotation device. E»S

hi-fi (HEE-fee), adj., f., high fidelity. E»S

highball (KHAI-bol), n., m., high-ball. Mixed alcoholic beverage. E»S Also JAIBO and HAIBOL.

hina (HEE-na), n., f., woman. "That hina is serious about Juan." {Ch, ELA}

hintear (HEEN-tear), v., to hint. E»S

hipertexto (hee-per-TEX-toh), n., m., hypertext. {CS} E»S

hiphop (HEEP-hop), n., m., rap music. E»S

hipi (HEE-pee), n., m./f., anti-establishment person. E»S Also HIPPY.

hipiteca (hee-pee-TE-kah), n., m., Aztec hippy. E»S

hippy (HEE-pee), n., m./f., anti-establishment person. E»S Also HIPI.

hit (HEET), n., m., hit. E»S

hitpareid (heet-pa-REID), n., m., hit parade. E»S

hits (HEETS), n., m., Internet hits. {CS} E»S

hoby (ho-BEE), n., m., hobby. E»S

hoja (OH-ja), n., f., corn husk used as cigarette wrapper. {SW}

hoky (ho-KEE), n., m., hokey. "That banda musical es hoky." E»S

hol (JOL), n., m., hallway. "Bret and Dulce first met en el hol." E»S

hollywoodiano (ho-le-woo-DEEA-no), n., m/f., Hollywood lifestyle. Sp. *Hollywoodense.* E»S

hombre (OM-bree), n., m., tough, mean. "He's not afraid. He's an hombre!" S»E

homlan (HOM-lan), n., f., homeland. "Vivimos lejos de nuestra homlan." E»S

homles (HOM-les), n., m., homeless. "Perdió el trabajo y since then es un homles." E»S

homo (HO-mo), n., m., 1. gay. 2. male person.

honestidad (ho-nest-TEE-dad), n., f., honesty. "Ella es una woman con honestidad." Sp. *honradez.*

honi (kho-NEE), n., m/f., 1. honey. 2. sweetheart. E»S Also HONY and JAINA.

honrón (hon-RON), n., m., homerun. Baseball term. "Mark McGwire hizo un honrón." ES

hony (kho‑NEE), n., m/f., 1. honey. 2. sweetheart. E»S Also HONI and JAINA.

hood (HUD), n., m., neighborhood. {NY, SW} E»S Also BARRIO and HUD.

hoodoo (HOO‑doo), n., m/f., 1. Jewish person. 2. evil incarnation. {Ch,} Also JEUDÍO.

hora feliz (Oh‑rah FE‑leez), n., f., happy hour. "La hora feliz en Mc‑Donald's is between 4 and 6 PM." E»S

hoste (HOS‑teh), n., m/f., host. "Mamá es la hoste de la casa." E»S

hotdog (hot‑DOG), n., m., sausage sandwich. "En el estadio de beisbol comemos hotdogs." E»S

hotline (hot‑LANE), n., f., telephone sex company. "Ella trabaja para una hotline in the evening." E»S

house (HAUS), n., m., type of music style. "Edmundo prefiere oir el house." E»S

hovercraft (hoh‑ver‑KRAFT), n., m., vehicle able to land on water. E»S Also HOVERFOIL.

huá! (GUAH), exp., Oh my! Sp. *gua!* "¡Huá!, she won the lottery."

huachiman (hua‑chee‑MAN), n., m., watchman. {SW} Also GUACHI‑MAN and WACHIMAN.

hud (HOOD), n., m., neighborhood. {SW} E»S Also BARRIO and HOOD.

huda (HU‑dah), n., m., police. "Javier tiene problemas con la huda." {SW}

huevos (HUE‑vos), n., m., testicles. "He was kicked en los huevos." Also, exp., *con huevos* means "with energy." {Ch, Mex}

huevos rancheros (HUE‑vos ran‑CHEE‑ros), n., m., Mexican break‑fast. "Yesterday we ate huevos rancheros at Friendly's." S»E

huisky (HUEES-kee), n., m., whisky. E»S

hula-hup (hoo-la-hoop), n., f., hoola-hoop. E»S

huligan (HOO-lee-gan), n., m., hooligan. E»S

humano (hu-MA-no), n., m/f., humanitarian. "Ella trabaja for a foun-dation que es humana." Sp. *humano* means human.

huntear (hun-TEHAR), v., to hunt. "La policía huntea al asesino." E»S Also HUNTIAR.

huntiar (khon-TEAR), v., to hunt. "La policía huntia al asesino." E»S Also HUNTEAR.

huntiar (khon-TEAR), v., to hunt. E»S Also HUNTEAR..

hurra (hoo-RRAH), exp., bravo. Eng. *hurrah*. E»S

husego (HOOS-go), n., m., jail. "Luego del crimen, John fue llevado al husego." Sp. *juzgado*. {SW}

husky (HOOS-key), n., m., as in clothes. {CA}

I

I.P.O. (EE‑poh), n., f., initial public offering. E»S

iceber (AYZ‑ber), n., m., iceberg. E»S

icefil (AYZ‑feel), n., m., polar field. E»S

ícon (EE‑kon), n., m., icon. {CS}

íglu (EE‑gloo), n., m., igloo.

ignición (ee‑GNEE‑tion), n., m., ignition. Sp. *máquina de encender.* E»S

iguana (ee‑GUA‑nah), n., f., arboreal lizard. The term dates from 1555. S»E

ilainer (ee‑LAI‑ner), n., m., e‑liner. {CS} E»S

imailiar (e‑MAI‑liar), v., to e‑mail. "Te voy a imailiar un mensaje esta noche de mi computador." {CS} E»S

imailito (e‑may‑LEE‑to), n., m., e‑mail. E»S Also E‑MAIL and MA‑NUELITO.

impacto (EEM‑pak‑toh), n., m., consequence. "El impacto of your action is a lawsuit." Sp. *impacto* means shock. {NY, SW} E»S

impechar (eem‑pe‑CHAR), v., to impeach. E»S

impechmen (eem‑PECH‑men), n., m., impeachment. E»S

implementar (eem‑pleh‑men‑TAR), v., to implement. E»S

implemento (eem‑pleh‑MEN‑toh), n., m., implement. "Este negocio necesita un implemento para mejorarse." E»S

impregnable (eem‑preg‑NAH‑ble), adv., impregnable. E»S

imprimir (eem‑pree‑MEER), v., to print. E»S

imputear (eem‑pooh‑TEAR), v., to offer input. E»S

in (EEN), exp., in style. E»S

inamistoso (en‑a‑mees‑TOH‑so), n., m/f., unfriendly.

incentivar (een‑cen‑tee‑VAR), v., to stimulate. "El jefe incentiva a sus empleados." E»S

incentivo (een‑cen‑TEE‑voh), n., m., incentive. Sp. *estimulante.* E»S

incluyendo (een‑cloo‑YEN‑doh), adj, included. "Todos viven en Nuyol, Martín incluyendo. E»S

incommunicado (een‑ko‑moo‑nee‑KA‑doh), n., and adj., m/f., incommunicated. E»S

inconsistencia (een‑kon‑sees‑TEN‑zeah), n., f., inconsistency. E»S

incumbente (een‑kum‑BEN‑teh), n., m., incumbent. E»S

incurrir (een‑koo‑REER), v., to incur. E»S

indentado (een‑den‑TAH‑do), adv., m/f., indented. "El texto quedó indentado." {CS} E»S

indentar (een‑den‑TAR), v., to indent. {CS} E»S

indiscernible (een‑dee‑cer‑NEE‑bleh), adv., undiscernible. E»S

indiscriminado (een-dees-kree-mee-NAH-do), n., m/f., indiscriminate. Eng. *indiscriminate.* E»S

indoctrinar (een-dok-TREE-nar), v., to indoctrinate. "Esos creyentes han sido indoctrinados por su iglesia." E»S

industrialista (een-dus-tria-LEES-ta), n., m., industrialist. E»S

ineluctable (een-eh-luk-TA-ble), adj., unavoidable. "Esta situación es ineluctable." E»S

inexhaustable (een-ex-aus-TAH-ble), adv., inexhaustible. E»S

infatuación (een-fa-tua-ZION), n., f., infatuation. Sp. *obsesión.* "Yo sufro de una infatuación con esa mujer." E»S

infatuado (een-fa-TUA-do), adj., m/f., infatuated. E»S

influenciado (een-flu-en-ZIA-do), n., m/f., influenced. "Javier Marías fue influenciado por Henry James." E»S

infuatuar (een-fa-TUAR), v., to infatuate. E»S

ingenuidad (een-ge-NUI-dad), n., f., 1. genius. 2. creativity. "La ingenuidad matemática de Isaiah es inmensa." Eng. *ingenuity.* E»S

ingenuo (een-GE-nuo), n., m/f., ingenuous. "That student es ingenuo. He always offers creative ideas." In Sp. *ingenuo* means näive. E»S

inglañol (en-gla-YNOL), n., m., mestizo language, part English, part Spanish, used predominantly in the United States since WWII. Also CASTEYANQUI, GRINGOÑOL and SPANGLISH.

inherente (een-eh-REN-teh), adv., inherent. "Ese problema social es inherente a la situación del mundo en general." E»S

inicializar (ee-nee-tiah-lee-ZAR), v., to initialize. "Santiago inicializa un nuevo documento." {CS} E»S

inin (EE-neen), n., m., baseball cycle. E»S

injunción (een-JUN-ktion), n., f., injunction. "A su tío le trajeron una injunción. Tuvo que salir de su apartamento." E»S

injuria (een‑JOO‑riah), n., f., injury. Sp. *injuria* means offense. E»S

input (EEN‑poot), n., m., input. E»S

insatisfactorio (een‑sa‑tees‑fak‑TO‑rio), n., m.f., unsatisfactory. E»S

inseminación (een‑she‑mee‑na‑ZION), n., impregnation. E»S

inseminado (een‑seh‑mee‑NAH‑do), n., m/f., impregnated. E»S

insensitivo (eren‑sen‑see‑TE‑vo), n., m/f., insensitive. E»S

inspectar (een‑spec‑TAR), v., to inspect. E»S

instalamento (een‑sta‑la‑MEN‑toh), n., m., installment. E»S

instrumental (een‑stroo‑MEN‑tal), adj., instrumental. E»S

insulado (een‑so‑LA‑do), adj., m/f., insulated. E»S

inteligencia (een‑telee‑GEN‑zia), n., f., secret police. E»S

interail (in‑TER‑rail), n., m., continental railway ticket. E»S

interestatal (een‑ter‑es‑ta‑TAL), adv. inter‑state. E»S

interfeiz (een‑ter‑FEIZ), n., m., interface. {CS} E»S

interferencia (een‑ter‑fe‑REN‑cia), n., f., electronic interference. E»S

interferir (een‑ter‑fe‑REER), v., to interfere. E»S

internacionalismo (een‑ter‑na‑zio‑na‑LEEZ‑mo), n., m., international‑ism. E»S

internet (een‑ter‑NET), n., m. international electronic network. "I found it navegando por el internet." E»S

internetear (een‑ter‑NEH‑tear), v., to browse the internet. {CS} E»S

interno (een‑TER‑no), n., m/f., intern. "El es un interno en el hospital." E»S

interviewstarse (een‑ter‑veeu‑STAR‑se), v., to interview onself. E»S

interviú (een‑ter‑VEEU), n., f., interview. Sp. *entrevista.* E»S

interviuiar (een‑ter‑VIW‑yar), v., to interview. E»S

introducción (een‑tro‑dook‑ZION), n., f., personal introduction. E»S

introducir (een-tro-doo-ZEER), v., to introduce people to one another. E»S

inversionista (een-ver-sio-NEEs-tah), n., m/f., investor. E»S Also IN-VESTOR.

investmengreid (een-ves-ment-GRAID), n., f., investment grade. E»S

investor (een-VES-tor), n., m/f., investor. "Quiero ser un investor de la bolsa." E»S Also INVERSIONISTA.

ir de campin (ir-de-KAM-pin), v. to go camping. "Fuimos de camping this semana pasada." E»S

irrelevancia (ee-reh-le-VAN-zia), n., f., irrelevance. E»S

irrelevante (ee-reh-le-VAN-teh), n., m/f., irrelevant. E»S

irreversible (ee-reh-ver-SEE-ble), n., m/f., irreversible. E»S

isis (EE-zee), adv., easy. E»S

isolismo (ee-zo-LEEZ-mo), n., m., isolationism. "El isolismo econó-mico shall increase the problem of poverty in that small coun-try." E»S

Istlos (EEST-los), n., m., East Los Angeles. {Ch, ELA, SW}

ítem (EE-tem), n., m., item. "Busco ese ítem en particular." E»S

itemizar (ee-the-mee-ZAR), v., to itemize. "Itemizamos los productors en la marqueta." E»S

J

jacal (kha·KAL), n., m., rude habitation. Literally *Indian hut.* "The peas-ants live in a jacal." S»E

jaibo (KHAI·bo), n., m., high·ball. Mixed alcoholic beverage. E»S Also HAIBOL and HIGHBALL.

jailero (khai·LE·roh), n., m/f., Hi·Lo tractor operator. {CA}

jailo (KHAI·lo), n., m., Hi·Lo tractor. {CA}

jaina (KHAI·na), n., f., 1. honey. 2. sweetheart. Also HONI and HONY.

jakear (HA·kear), v., to hack. {CS} E»S Also JAQUEAR and HA-KEAR.

jaker (HA·ker), n., m/f., computer hacker. {CS} E»S Also HA-QUER and HAKER.

jaket (JA·ket), n., m. jacket. "It's cold outside, so llévate tu jaket." E»S Also JAQUET.

jalapeño pusher (kha·la·PE·nyo POO·sher), n., m., petty criminal. E»S

jale (KHA·le), n., m., job. "Abraham perdió su jale en la lunchería." {CA}

jandimán (khan·dee·MAN), n., m., handyman. E»S Also HANDI·MAN.

janguear (KHAN·gehar), v., 1. to hang out. 2. to loiter. "Please Papi, solo vamos a janguear en la casa del Flaco." E»S Also HAN·GAR, HANGEAR, HANGUIAR and JANGUIAR.

janguiar (KHAN·gehar), v., 1. to hang out. 2. to loiter. "Please Papi, solo vamos a hangear en la casa del Flaco." E»S Also HANGAR, HANGEAR, HANGUIAR and JANGUEAR.

jaquear (HA·kear), v., to hack. {CS} E»S Also HAKEAR and JAKEAR.

jaquet (JA·ket), n., m. jacket. "It's cold outside, so llévate tu jaquet." E»S Also JAKET.

javear (kha·VEHAR), v., to have. "Ella java un automóvil green." Sp. *tener* means to have. E»S Also HAVEAR, JAVEAR and JA·VEAR.

javiar (kha·VEHAR), v., to have. "Ella javia un automóvil green." Sp. *tener* means to have. E»S Also HAVEAR, HAVIAR and JAVIAR.

jaz (JAZ), n., m., jazz. "He went al Blue Note a escuchar una banda de jaz."

jazband (jaz·BAND), n., f., Jazz troupe. "Fuimos a escuchar la jaz·band de Nueva Orleans."

jazeando (ja·ZEAN·do), adv., act of improvising music or another in·tellectual or artistic activity. E»S

jazear (ja·ZEAR), v., to improvise. E»S

jazero (ja·ZE·ro), n., m/f., 1. improviser. 2. jazz specialist. E»S

jeans (JEENS) n., m., jeans pants. Sp. *pantalón de mezclilla.* E»S Also JINS, LEVIS and LIVAIS.

jeep (JEEP), n., m., sports automobile with capacity to navigate rough terrain. E»S Also JIP.

jefa (KHE-fa), n., m., 1. mother, maternal figure. 2. female boss, supervisor. {Ch, Mex, SW}

jefe (KHE-fe), n., m., 1. father, paternal figure. 2. male boss, supervisor.

jefita (khe-FEE-tah), exp., endearing term for mother. "I can't go because my jefita will get mad." {Ch}

jerga (KHER-ga), n., f., 1. heavy woolen fabric used for floor covering. 2. jargon.

jerki (JER-kee), n., m/f., jerk. "Antonio se comporta como un jerki con los jórones." Possible Peruvian origin. Also from American Indian *charqui.* {SW}

jersey (JER-see), n., m., 1. T-shirt, jersey; 2. New Jersey. Also JERSISITI. "Dad, can you buy me ese jersey de Pato Donald, por favor?" E»S

Jersisiti (jer-see SEE-teee), n., Jersey City. "Pasamos las vacaciones a Jersisiti this summer."

jet (JET), n., m., fast-speed airplane. E»S Also YET.

jetfoil (JET-foil), n., m., hovercraft. E»S

jetlag (JET-lag), n., m., jet lag. "Viajé a Berlín y tengo un jetlag." E»S

jetset (JET-set), n., m., trend-setting elite or style. "Enrique Iglesias es parte del jetset." E»S

jeudío (kheu-DEE-o), n., m/f., 1. Jewish person. 2. evil incarnation. {NM} Also HOODOO.

jinete (khee-NE-teh), n., m/f., first-class horse-rider. S»E Also KOQUI, and JOKI.

jingo (JEEN-go), n., m., sound. "En la noche oíamos un jingo outside the house."

jins (JEENS), n., m., jeans pants. Sp. *pantalón de mezclilla*. E»S Also JEANS, LEVIS and LIVAIS.

jip (JEEP), n., m., sports automobile with capacity to navigate rough terrain. E»S Also JEEP.

jobi (KHO-bee), n., m., hobby, pass-time. E»S

jogin (JOO-geen) v., jogging. E»S Also FOOTIN, FUTIN, FU-TINEAR and FUTINIAR.

johnsonear (jon-so-NEHAR), v., 1. to reflect on linguistic issues. 2. to apply one's intellect to the understanding of literature. E»S {Mas} Also JOHNSONIAR.

johnsoniar (jon-so-NEHAR), v., 1. to reflect on linguistic issues. 2. to apply one's intellect to the understanding of literature. E»S {Mas} Also JOHNSONEAR.

joistic (JOY-steek), n., m., 1. candy food. 2. Joystick, computer term. {CS} E»S

joker (JOE-ker), n., m., the extra playing card without fixed value, sometimes used as the highest trump or as wild card. "El ganador del juego finishes with the joker in his hand." E»S

joki (KHO-kee), n., m/f., first-class horse-rider. E»S Also JINETE and JOQUI.

jol (KHOL), n., m., hallway, pathway. E»S

joldar (khol-DAR), v., 1. to hold up. Reference to phone device. 2. to hold, to sustain. E»S Also JOLDEAR and JOLDIAR.

joldear (khol-DE-ar), 1. to hold up. Reference to telephone device. 2. to hold, to sustain. E»S Also JOLDAR and JOLDIAR.

joldero (khol-DE-ro), n., m/f., person that is on-hold in a phone conversation. E»S

joldiar (khol·DEHAR), v., 1. to hold up. Reference to phone device. 2. to hold, to sustain. E»S Also JOLDAR and JOLDEAR.

jondipear (khon·dee·PEHAR), v., to buy hardware tools. E»S

jondípo (khon·DEE·po), n., m., hardware store. After brand name Home Depot. {CA} E»S

jonrón (khon·RON), n., m., home run. Baseball term. "Manny Ramirez metió un jonrón." E»S

joqui (KHO·kee), n., m/f., first·class horse·rider. E»S Also JINETE, JOQUI and KOQUI.

jornada (khor·NA·da), n., f., full day's workload. S»E

jornada·del·muerte (khor·na·da·del·MUER·teh), n., f., waterless land. {SW}

jornalero (khor·na·LEH·ro), n., m/f., a person with a nine·to·five job.

joyceano (joy·ZEEA·no), n., m/f., joyful. E»S

jugarpape (kho·gar·PA·pe), v., to play a dramatic role. "Nicanor juega· pape a Rosencrantz en la obra *Hamlet*." {CA}

jumbo (JUM·bo) n., m., jumbo jet. E»S

jumpear (jum·PEAR), v., to jump. E»S

jumper (JUM·per), n., m., 1. sweater. 2. person who jumps. E»S

jungla (KHOON·gla), n., f., 1. metropolis. 2. crime·infested section of town. "Muchos millones viven en la jungla de Nuyol." E»S

junior (JOO·nyor) n., m., 1. wealthy young man. 2. Young person with political and financial connection. From Latin *juvenior*. "Ese ju· nior todavía tiene que aprender mucho." E»S Also YUNIER and YUNIOR.

junkería (khoon·ker·REE·a) n., f., junk store. "Me gustan las junkerías porque they always have interesting trash." E»S Also JUN· QUERIA.

junquería (khoon‑ker‑REE‑a) n., f., junk store. "Me gustan las jun‑querías porque they always have interesting trash." E»S Also JUNKERIA.

junta (KHOON‑ta), n., f., 1. committee meeting. "The comité is in a junta." 2. military group in power. The term dates from 1623. S»E

justiciable (khoos‑tee‑ZEEA‑ble), adv., justifiable. "Ese error es justi‑ciable." Sp. *justificable.* E»S

justificar (khoos‑tee‑fee‑KAR), v., to justify. "El justifica his actions de forma unconvincing." E»S

justinta (khoos‑TEEN‑na), exp., punctual, on time. Eng. *just in time.* "She arrive justinta al concierto." E»S

justo (KHOOS‑toh), adj., m/f., honest. "Vicente es un hombre justo. He never steals any money." Sp. *honesto.* E»S

K

kartin (KAR‑teen), n., m., car race. E»S

kartón (kar‑TON), n., m., cardboard. {NY, SW} E»S

kayac (KA‑yak), n., m., kayak.

kechar (ke‑CHAR), v., to catch. "Ni kecha, ni pitcha, ni deja batear." Also CACHAR.

keis (KEIS), n., m., case. 1. legal case. "El abogado works en el keis de Juanito." 2. exp., in case. {Ch} Also CEIS and QUEIS.

kennedito (ke‑ne‑DEE‑to), n., m., traitor. Reference to John F. Kennedy's involvement in Bay of Pigs. "Armando es un kennedito. Nobody ought to trust him." {CA} E»S

ketcho (KE‑cho), n, m., ketchup. "Las hamburguesas no saben bien sin ketcho." E»S

kicheneta (kee‑che‑NE‑ta), n., f., kitchenette. E»S

kikeado (kee‑KEA‑doh), n., m/f., kicked around by the police. {ELA} E»S

kilkar (KEEL-kar), v., to kick around. {ELA} E»S

kilobait (kee-lo-BAIT), n., m., kilobyte.

kilt (KEELT), n., f., Scottish dress used by males. "El escocés usa su kilt." E»S

kit (KEET), n., m., medical kit. "El doctor siempre lleva un kit, en caso de accidente." E»S

Kiues (KEE-wes), n., Key West. E»S

kiut (KEE-ut), n., m/f., cute. E»S

klinex (KLEE-nex), n., m., tissue paper. After the brand name. "¿Me regala un klinex, plis?" Also CLINE

knocaut (KNOK-aut), n., m., knock out. E»S

knowhau (KNOU-hau), n., m., itemized knowledge. "I need the knowhau in order to arrive to your house." E»S

kodak (KOO-dak), n., f., 1. camera. 2. photograph. After brand name. "Take a kodak of the children while they swim."

koqui (KHO-kee), n., m/f., first-class horse-rider. Also JINETE, JOKI and JOQUI.

krijma (KREEJ-mah), n., f., Christmas. Used in place of *Navidades*. {CA} E»S Also CRISMAS.

kril (KRIL), n., m., shrimplike crustaceans.

kuriosita (koo-rio-SEE-tah), adj., f., cute girl. "That girl está bien kuriosita." {ELA}

L

labor (la-BOR), n., f., activity, employment. "El tiene una labor de plomero." E»S

laborismo (la-bo-REES-mo), n., m., activities related to work. E»S

laca (LA-kah), n., f., door lock. The term often refers to automobiles. "No se te olvide ponerle la laca a the door of the car." {PR}

lactoso (lak-TO-so), n., m., 1. dairy-related product. 2. period in a baby's life in which milk is the sole nutrient. 3. milk-drinking person, usually an infant. 4. suffering of allergy to lactose products.

ladina (la-DEE-na), 1. n., f., wild cow. 2. n., f., savage animal. 3. adj. crafty. {Ch}

lady (LAY-dee), n., f., respectful appellation to a woman. 2. distinguished female. E»S

lager (LA-ger), n., f., light beer. After brand name. E»S

laguear (la-GEHAR), v., to lag behind. {CS} E»S

lait (LAIT), 1. adj., smooth. 2. suave. 3. calm. 4. n., f., electric light. Also LIGHT and LITE.

lambswool (lambs-WOOL), n., m., type of fabric. E»S

lando (LAN-do), n., m., landlord. E»S Also LANLOR and LANLO.

landrover (lan-RO-ver), n., m., automobile designed for rough terrain.

lanlaidi (LAN-ley-dee), n., f., female landlord. E»S

lanlo (LAN-lo), n., m., landlord. E»S Also LANLOR and LANDO.

lanlor (LAN-lor), n., m., landlord. E»S Also LANLO and LANDO.

lanolina (la-no-LEE-na), n., f., lanoline.

laptop (LAP-top), n., f., portable computer. "Me llevé la laptop to the trip." {CS} E»S Also LAPTOPA.

laptopa (lap-TO-pah), n., f., portable computer. "Me llevé la laptopa to the trip." {CS} E»S Also LAPTOP.

lasso (LA-zo) n., m., long-throwing rope with a running noose at one end; originally from the Spanish *lazo*. The terms date from 1768. S»E.

látigo (LA-tee-go), n., m., strap used to cinch a saddle. S»E

latin-lover (la-teen-LO-ver), n., m., 1. passionate lover. 2. attractive male. "Enrique Iglesias es un latin-lover." E»S

laudspiquer (LAUD-spee-ker), n., m., loud speaker. E»S

lavatorio (la-va-TO-rio), n., m., rest room.

lave ho! lave! (lav-KHO-lav), exp., 1. mountain person. 2. exp., wake up! Sp. *levantar.* E»S

lead (LEED), n., m., 1. leading item. 2. headline. 2. leading piece of news. "El lead de esta mañana anunció la muerte de Tito Puente." E»S Also LID.

leanear (lee·NEHAR), v., to lean. E»S Also LEANIAR, LINEAR and LINIAR.

leaniar (lee·NEAR), v., to lean. E»S Also LEANEAR, LINEAR and LINIAR.

leasin (LEE·seen), n., m., 1. lease. 2. Act of renting an item of real estate. E»S Also LIS.

lectura (lek·TOO·ra), conference. Sp. *conferencia*. E»S

ledear (lee·DEAR), v., to lead. E»S Also LIDEAR.

leggin (LE·geen), n., m., type of stretch pants. Eng. *legging*. E»S Also LEGIN.

leguín (le·GEEN), n., m., type of stretch pants. Eng. *legging*. E»S Also LEGUI.

lépero (LE·pe·roh), n., m., 1. thief. 2. fool·mouthed person.

leva (LE·va), n., m., loser. "That guy has always been a leva." In Mex. Sp. *le vale madre*. {ELA}

levis (LE·vis) n., m., jeans pants. After Levi's brand name. Sp. *pantalón de mezclilla*. Also JEANS, JINS and LIVAIS.

lexicópata (le·xee·KO·pa·ta), n., m., a harmless drudge. {Mas}

leyof (LAY·of), n., m., layoff. A Juan le dieron un leyof yesterday." E»S

leyofear (le·yo·FEHAR), v., to be laid off. E»S Also LEYOFIAR.

leyofiar (le·yo·FEHAR), v., to be laid off. E»S Also LEYOFEAR.

libor (LEE·bor), n., m., acronym for London Interbanking Offered Rate. Price of money or type of basic interest in the interbanking market in London.

librería (lee·bre·REE·a), n., f., library. In Sp. *biblioteca* is library and *librería* is bookstore. E»S

lid (LEED), n., m., 1. leading item. 2. headline. 2. leading piece of news. E»S Also LID.

lidear (lee-DEHAR), v., to lead. In Sp. *lidear* means to confront. E»S Also LIDERAR and LIDEREAR.

líder (LEE-der), n., m/f., leader.

liderear (lee-de-REHAR), v., to lead. In Sp. *lidear* means to confront. E»S Also LIDEAR and LIDERIAR.

lideriar (lee-der-REHAR), v., to lead. In Sp. *lidear* means to confront. E»S Also LIDEAR and LIDEREAR.

lift (LEAFT), n., m., 1. automobile ride. 2. occasional help. E»S

lifteador (leaf-tea-DOR), n., m/f., person in charge of lifting items. E»S Also LIFTIADOR.

liftear (LEAF-tear), v. to lift. "No puede liftear the box." E»S Also LIF-TIAR.

liftiador (leaf-tea-DOR), n., m/f., person in charge of lifting items. E»S Also LIFTEADOR.

liftiar (LEAF-tehar), v. to lift. "No puede liftiar the box." E»S Also LIFTEAR.

liftin (LEAF-teen), n., m., 1. lift. 2. pick-up. 3. plastic-surgery operation to eliminate wrinkled skin. E»S

light (LAIT), adj., 1. smooth. 2. suave. 3. calm. 4. n., f., electric light. E»S Also LAIT and LITE.

lik (LEEK), n., m., leak. {CA} E»S

lima (LEE-ma), n., f., lemon.

liminal (lee-mee-NAL), n., m., basement. From Latin *liminal*. "Ella vive en el liminal de su bildin." Sp. *subterráneo*. Also BEISMEN.

limón (lee-MON), n., m., lime.

limusina (lee-moo-SEE-na), n., f., limousine.

lina (LEE-na), n., f., line of cocaine.

linchar (leen-TCHAR), v., to lynch. E»S

linear (lee-NEHAR), v., to lean. Sp. *linear* means linear. E»S Also LEA-NEAR, LEANIAR and LINIAR.

liniar (lee-NEHAR), v., to lean. "Ella se linia en su madre para pagar the rent." E»S Also LEANEAR, LEANIAR and LINEAR.

link (LINK), n., m., Internet connection. {CS} E»S Also LINQUE.

linkear (leen-KEAR), v., to link. {CS} E»S Also LINKIAR, LINQUEAR and LINQUIAR.

linkiar (leen-KEAR), v., to link. {CS} E»S Also LINKEAR, LINQUEAR and LINQUIAR.

linóleo (lee-NO-le-o), n., m., linoleum. From Latin. "Alison puso linoleo en la cocina."

linotipia (lee-no-TEE-peha), n., f., linotype.

linque (LINK), n., m., Internet connection. {CS} E»S Also LINK.

linquear (leen-KEAR), v., to link. {CS} E»S Also LINKEAR, LINKIAR, and LINQUIAR.

linquiar (leen-KEAR), v., to link. {CS} E»S Also LINKEAR, LINKIAR, and LINQUEAR.

lipistiquear (leep-stee-KEAR), v., to put on lipstick. E»S Also LIPISTIQUIAR.

lipistiquiar (leep-stee-KEAR), v., to put on lipstick. E»S Also LIPISTIQUEAR.

lipstic (LEEP-stik), n., m., lipstick.

liqueando (lee-ke-AN-do), v., to slobber with the tongue. "Estaba lickeando la paleta, even after it fell down on the floor." E»S

liquear (lee-KIAR), v. to leak. "La lavamanos está liqueando desde esta mañana." E»S Also LIQUIAR

liqueo (lee-KE-o), n., m., the act of leaking. {CA} E»S

liquiar (lee-KEAR), v., to leak. E»S Also LIQUEAR.

lis (LEES), n., f., lease. E»S Also LEASIN.

lisa (LEE-sa), n., f., shirt. {Ch}

lisear (lee-SEAR), v., to lease, to rent. E»S Also LISIAR.

lisiar (lee-SEAR), v., to lease, to rent. E»S Also LISEAR.

lista-caliente (lees-ta ka-LEE-en-te), n., f., hot list.

listin (LEES-teen), n., m., classified list. E»S

lite (LAIT), 1. adj., smooth. 2. suave. 3. calm. 4. n., f., electric light. Also LAIT and LIGHT.

líter (LEE-ter), n., m., trash. "Acumulamos mucho líter despues de la fiesta."

literacía (lee-te-ra-SEE-), n., f., reading knowledge, reading habit.

literalmente (lee-te-RAL-men-te), adv., literally.

litoset (lee-to-SET), n., m., offset. E»S

liváis (LEE-vays) n., m., jeans pants. After Levi's brand name. In Sp. *pantalón de mezclilla.* Also JEANS, JINS and LIVAIS.

liverant (LEE-ve-rant), n., m., monster. {Mex}

livin (LEE-been), n., m., 1. living room. "Mi cuarto favorito es el livin." 2. income. "Trabajo mucho pero no hago el livin suficiente." E»S

llamar pa'tras (ya-mar pa-TRAS), exp., to return a phone call. Eng. *to call back.* Sp. *regresar la llamada.* "Llámalo pa'tras por el celiular!"

llano (YAH-no) n., m., flat, open prairie. "El llano siempre ha sido un sitio preferido pa' ser campin."

llegue (YE-gue), n., m., automobile crash. "Sufrió un llegue en su Volvo." {Ch, Mex}

llorona (yo-RO-na), n., f., troubled soul. From the myth of *La Llorona.* {SW}

loadeador (lo-dea-DOR), n., m/f., loader {CS} E»S

loadear (lo-DEHAR), v., to load. {CS} E»S

lob (LOB), n., m., swing ball that passes on top of the adversary. Tennis term.

lobear (lo-BEAR), v., to put pressure politically. Also LOBIAR.

lobi (LO-bee), n., m., 1. lobby. 2. political group. E»S

lobiar (lo-BEAR), v., to put pressure politically. E»S Also LOBEAR.

lobista (lo-BEES-ta), n., m/f., person involved in a political group. E»S

lobo (LO-bo) n., m., sexual abuser. "El fue acusado de lobo by the secretary."

locaut (lo-KAUT), n., m., 1. company closing. 2. massive firing of workers in response to a strike. E»S

loco (LO-ko), adj., insane, crazy. The term was adopted in English around 1930. S»E

loco-da-poco (lo-ko da PO-koh) exp., temporarily out of Control. "Silvio saw the girl and became loco-da-poco."

locomotor (lo-ko-mo-TOR), n., m., locomotive. Also LOCOMO-TORA.

locomotora (lo-ko-mo-TO-ra), n., m., locomotive. Also LOCOMOTOR.

loft (LOFT), n., m., open-space apartment. "Compró un loft en Nueva York."

logear (lo-GEHAR), v., to log in. {CS}

Loisiada (loy-SEEHA-da), n., Lower East Side. {NY} Also LOSAIDA.

loma (LO-mah), n., f., broad hill. "Our town is built on a loma." The term dates from 1864. S»E

lonch (LONCH), n., m., 1. mid-day meal. 2. food served to guests at an event. "Hijito, no se te olivide tu lonch today." E»S Also LUNCH and LONCHE.

lonchar (lon-TCHEAR), v., to lunch. E»S Also LONCHEAR and LONCHIAR.

lonche (LONCHE), n., m., 1. mid-day meal. 2. food served to guests at an event. "Hijito, no se te olivide tu lonche today." E»S Also LUNCH and LONCH.

lonchear (lon-TCHEAR), v., to lunch. E»S Also LONCHAR and LONCHIAR.

lonchería (lon-tche-REE-ah), n., f., cafeteria, lunch-serving restaurant. E»S

lonchiar (lon-TCHEAR), v., to lunch. E»S Also LONCHAR and LONCHEAR.

londri (LON-dree), n., f., laundry.

lonplei (lon-PLEY), n., m., 33^1/$_3$-revolution music record. Sp. *elepé*.

look (LOOK), exp., style, exterior image, aspect. "The actress tiene un look especial." Also LUK.

loopear (loo-PEAR), v., to loop. E»S Also LUPEAR.

loopin (LOO-peen), n., m., acrobatics in which a person performs a complete vertical circle in the air. E»S Also LUPIN.

lor (LOR), n., m., boss, landlord. "We owe $875 al lor por la renta of this month." E»S

Losaida (loy-SAEE-da), n., Lower East Side. (NY, PR) Also LOI-SIADA.

luk (LOOK), exp., style, exterior image, aspect. "The actress tiene un look especial." Also LOOK.

luminaria (loo-mee-NA-ria), n., f., 1. Christmas light. 2. celebrity.

luminífero (loo-mee-NEE-phe-roh), adj., luminous. E»S

lunch (LONCH), n., m., 1. mid-day meal. 2. food served to guests at an event. "Hijito, no se te olivide tu lunch today." E»S Also LONCH and LONCHE.

lupear (loo-PEAR), v., to loop. E»S Also LOOPEAR.

lupin (LOO-peen), n., m., acrobatics in which a person performs a complete vertical circle in the air. E»S Also LOOPIN.

M

macadam (mah‑KA‑dam), n., m., seducer of women. "Don't be such a macadam." {NY} Also DON JUAN and MACADAMO.

macadamo (mah‑ka‑DA‑mo), n., m., seducer of women. "Don't be such a macadamo." {NY} Also DON JUAN and MACADAM.

macademizar (ma‑ka‑de‑mi‑ZAR), v., to seduce.

mach (MATCH), n., m., encounter. Usually used in athletic activities to denote a contest. E»S

machar (ma‑TCHAR), v., to match. E»S Also MACHEAR.

machear (ma‑TCHEAR), v., to match. E»S Also MACHAR.

macheo (ma‑TCHE‑o), n., m., match‑up. "Ella necesita un macheo con alguien who could understand her." E»S

machismo (ma‑CHEES‑mo), n., m., 1. exaggerated sense of masculinity. 2. masculine pride. S»E

macho (MA-tcho), n. and adj., Particularly masculine, aggressively vir-
ile. "Male bravado" Sp. *macho* means *male.*

machpoint (match-POINT), n., m., last match of a game. Tennis term.
"Pete Sampras ganó el machpoint contra Andre Agassi." E»S

macutear (ma-KU-tear), v., to chat. {D}

macuteo (ma-ku-TE-o), n., m., act of chatting. {D}

madama (ma-DA-ma), n., f., female pimp. Eng. *Madam.*

magazín (ma-ga-ZEEN), n., f., magazine. E»S

maicro (MAY-kro), n., m., microwave oven. E»S Also MAICRO-
GÜEY.

maicrogüey (mee-kiro-WEY), n., m., microwave oven. E»S Also MAI-
CRO.

maicup (MEY-kup), exp., m., cosmetics. E»S

maidin (MEID-in), n., m., made in. "Esta TV es maidin en Japan."
E»S

mailear (MAY-lear), v., to mail. "Te maileo mañana." E»S

maileo (may-LE-o), n., m., act of mailing a letter. E»S

mailin (MAY-leen), n., m., items or information being mailed out. E»S

maleta (ma-LE-ta), n., f., satchel. {Ch, Mex}

malinchi (ma-LEEN-tchi), n., m/f., an unpatriotic Mexican. From *mal-
inchista.* {Ch, Mex}

Mallamibish (ma-ya-mi-BISH), n., Miami Beach. {CA}

malpaís (mal-pa-EEZ), n., m., region of lava flats in the southwestern US.

malta (MAL-ta), n., f., malt.

manache (ma-NA-tche), n., m/f., manager, administrator. {CA} E»S
Also MANAGER and MANAYER.

manachear (ma-na-TCHEAR), v., to manage, to administer. E»S

manada (ma-NA-da), n., f., herd of mares.

manager (ma-na-DGER), n., m., manager, administrator. "María es la manager de la corporación." E»S Also MANACHE and MA-NAYER.

management (ma-nadg-MENT) n., m., management. "Rosario es parte del management de la empresa." E»S

mañana (ma-NYA-nah), n., m/f., tomorrow, sometime, maybe never. Sp. *mañana* means morning and tomorrow.

manayer (ma-na-YER), n., m., manager, administrator. E»S Also MA-NACHE and MANAGER.

mancito (man-SEE-to), n., m., youngster. "Ese mancito es conocido for getting himself in fights." {NY}

mandatorio (man-da-TOH-rio), adj., m/f., mandatory. "Esta actividad es mandatorio." E»S

mandril (man-DREEL), n., m., mandrill.

mangana (man-GA-na), n., f., type of lasso grabbed by the forefeet.

mango (MAN-go)., n., m., 1. tropical fruit. 2. handsome male. "Antonio Banderas es un mango."

mangua (MAN-gua), n., m., man of war. {Chi}

maniquiú (ma-nee-CUE), n., m., manicure. "Liora se hace un maniquiú." In Sp. *hacerse la manicuta.*

maniquiurista (ma-nee-cu-REES-ta), n., m/f., manicurist. Also MANI-QUIRI.

maniquiri (ma-nee-KEE-ree), n., m/f., manicurist. Also MANIQUIU-RISTA.

mantilla (man-TEE-ya), n., f., female veil. The term dates from 1717. S»E

manuelito (ma-nwe-LEE-to), n., m., e-mail. E»S Also E-MAIL and IMAILITO.

mapear (ma-PEAR), v., 1. to map. "Pedro mapea la ruta de Nueva York a Boston." 2. to mop. "Make sure to mapear the kitchen." E»S Also MAPIAR, MOPEAR and MOPIAR.

mapiar (ma-PEAR), v., 1. to map. "Pedro mapia la ruta de Nueva York a Boston." 2. to mop. "Make sure to mapiar the kitchen." E»S Also MAPEAR, MOPEAR and MOPIAR.

maraja (ma-RAH-kha), n., f., maharajah, sovereign of an Indian province.

mari (MA-ree), n., f., marijuana. Also MARY.

marielito (ma-rie-LEE-to), n., m/f., Cuban exile who left the homeland during the Mariel Boatlift in 1980.

marín (ma-REEN), n., m., marine, person who is part of the navy. E»S Also MARINO.

marino (ma-REE-no), n., m., marine, person who is part of the navy. Also MARIN.

mariposa (ma-ree-PO-sa), n., f., homosexual. {Co}

marketín (mar-ke-TEEN), n., m., marketing, organization and distribution of products. E»S Also MARKETING and MARQUETIN.

marketín directo (mar-ke-teen dee-REC-to), n., m., 1. marketing without intermediaries. 2. direct-mail publicity. E»S

marketing (mar-ke-TING), n., m., marketing, organization and distribution of products. E»S Also MARKETIN and MARQUETIN.

marketín mix (mar-ke-TEEN MIX), n., m., a combination of different commercial methods and instruments used by a company to accomplish a fixed objective. E»S

marketsher (mar‑ket‑SHER), n., m., a share in the stock market. "Tengo un marketsher on General Electric." E»S

marqueta (mar‑KE‑tah), n., f., supermarket. E»S Also SUPER.

marquetín (mar‑ke‑TEEN), n., m., marketing, organization and distribution of products. E»S Also MARKETIN and MARKETING.

mary (MA‑ree), n., f., marijuana. Also MARI.

mascara (mas‑KA‑rah), n., f., makeup.

maskimteip (mas‑keem‑TEIP), n., f., glued tape. E»S

masmedia (mas‑ME‑dya), n., f., mass media. E»S

mastear (mas‑TEAR), v., to master. E»S

máster (MAS‑ter), n., m., 1. master's degree. 2. master person in a specific activity. "Gustavo es un master en soccer."

matador (ma‑ta‑DOOR), n., m., bullfighter. The term dates from 1681. S»E

materializar (ma‑te‑rea‑lee‑CZAR), v., to materialize. "El amigo misterioso se materializó en la escuela a las 12 PM." E»S

matrasa (ma‑TRA‑sah), n., f., mattress. E»S

matrasear (ma‑tra‑SEAR), v., 1. to rest. 2. to sleep. E»S

mauntainbaic (mawn‑tain‑BAIK), n., m., athletic activity in which bycles are used for climbing steep terrain. E»S

maus (MAUS), n., m., computer mouse. "Mueve el maus para abrir la window." {CS} E»S Also MOUS and MOUSE.

maximalista (ma‑xee‑ma‑LEES‑ta), n., m/f., self‑aggrandizer. E»S

maximilización (ma‑xee‑mee‑lee‑za‑ZION), n., act of maximizing an item or pattern. E»S

maximizar (ma‑xee‑mee‑CZAR), v., to maximize. E»S

maxisingle (ma‑xee‑SEEN‑gle), n., m., music disk played at 45 revolutions per minute. E»S

mayor (ma-YOR), n., m., major. In Sp. *alcalde* means major. Also, *mayor* means biggest. E»S

mazorra (ma-ZOO-rra), n., f., psychiatric hospital. {CA}

meanin (MEE-neen), n., m., meaning. E»S Also MININ.

mecha (ME-tcha), n., f., lighting match. "Tienes una mecha pa' mi cigarro? E»S

Mediquei (me-dee-KEY), n., m., government-subsidized healthcare system. E»S Also MEDIQUEI.

Mediquer (me-dee-KERE), n., m., government-subsidized healthcare system. E»S Also MEDIQUEI.

megabait (me-ga-BAIT), n., m., one million bytes of memory stored in a computer system. E»S

megatón (me-ga-TON), n., m., units of measurement of energy in a nuclear bomb. From the Greek.

meikear (may-KEAR), v., to succeed, to triumph. Often preceded by the pronoun *la*. "Juana la meikea en Estados Unidos." E»S Also MEIQUEAR, MEIQUIAR and MEIKIAR.

meikiar (may-KEAR), v., to succeed, to triumph. Often preceded by the pronoun *la*. "La meikié en América." Also MEIQUEAR, MEIQUIAR and MEIKEAR.

meikit (MEY-kit), exp., Make it! "Meikit, Rodrigo. You have the talent!"

meiquear (may-KEAR), v., to succeed, to triumph. Often preceded by the pronoun *la*. "La meiquié en América." Also MEIQUIAR, MEIKEAR and MEIKIAR.

meiquiar (may-KEAR), v., to succeed, to triumph. Often preceded by the pronoun *la*. "La meiquí en América." E»S Also MEIQUEAR, MEIKEAR and MEIKIAR.

membersía (mem-ber-SEE-a), n., f., membership. "She has a member-sía para el club de rotarios."

mentirita (men‑tee‑REE‑ta), n., f., alcoholic mix usually made of rum and Coke. Also known as *Cuba Libre.* "I ordered a mentirita at the bar." {CA}

meraneta (mera‑NE‑ta), exp., the sole truth. "Jaime said the meraneta about the incident." {Ch, Mex}

mercadista (mer‑ka‑DEES‑ta), n., m/f., person involved in business transactions.

mercadizar (mer‑ka‑dee‑ZAR), v., to turn an item into merchandize.

mercandaisin (mer‑kan‑DAI‑seen), n., m., promotion and commercialization of a product.

mercantbank (mer‑kan‑BANK), n., m., merchant's bank.

mergear (MER‑ger), n., m., merger. E»S Also MERGER.

merger (MER‑ger), n., m., merger. E»S Also MERGEAR.

mergerizar (mer‑ge‑ree‑CZAR), v., to merge. Also used in the reflexive form. "Una compañía se mergeriza con la otra." E»S

mestizo (mez‑TEE‑soh), n., m/f., person of mixed race. The term dates from 1582. S»E

Mexkimo (mex‑KEE‑mo), n., m/f., Mexican in or from Alaska. {SW}

mica (MEE‑ka), n., f., green card, working papers. E»S

michiman (MEE‑chee‑man), n., m., midshipman. E»S

microchip (mee‑kro‑CHEEP), n., m., minuscule computer chip. {CS} E»S

microfilme (mee‑kro‑FILM), n., m., microfilm. E»S

microfish (mee‑kro‑FISH), n., m., microfiche. E»S

migra (MEE‑gra), n., f., immigration police. Also used to refer to the Immigration and Naturalization Service [INS] staff. "La migra de los gringos is after us."

milaidy (mee-LAY-dee), n., f., my lady. E»S

míldu (MEEL-doo), n., m., mildew. E»S

milord (mee-LORD), n., m., my lord. E»S

mini (MEE-nee), n., f., short skirt. "A María le gustan las minis." E»S

minibasket (mi-ni-BAS-ket), n., f., kid basketball, practiced on a smaller court than usual, with lower baskets. E»S

minigolf (mee-nee-GOLF), n., m., small golf course. E»S

minimización (mee-nee-mee-za-ZION) n., f., reduction of an item or condition to its minimum potential. E»S

minín (mee-NEEN), n., m., meaning. E»S Also MEANIN.

minoridad (mee-no-ree-DAD), n., f., minority. Sp. *minoría*. E»S

minuta (mee-NOO-ta), n., f., detail. "La minuta en ese libro is incredible." Eng. *minutia*. E»S

mis (MEES), n., m., beauty queen. "Mis Venezuela." E»S Also MISS.

misear (mee-SEE-ar), v., to miss, to long for. Also MISIAR.

misi (MEE-see), n., f., missus. "Misi Johnson era mi schoolteacher." E»S

misiar (mee-SEE-ar), v., to miss, to long for. E»S Also MISEAR.

misil (me-SEAL), n., m., missile. "En la batalla de Bagdad se usaron misiles balísticos."

misin (MEE-ssen), n., m/f., missing person. E»S Also MISING.

mising (MEE-ssen), n., m/f., missing person. E»S Also MISIN.

misionario (mee-zio-NA-rio), n., m/f., missionary.

miss (MEES), n., m., beauty queen. "Ella es la nueva Miss Venezuela." E»S Also MIS.

mistear (mees-TEAR), v., to miss. "Te misteo desde que te fuistes a San Francisco." E»S

míster (MEES‑ter), n., m., 1. mister. 2. boss. 3. acquaintance. 4. athletic trainer. 5. gringo. Originally developed in Puerto Rico in the period of mandatory English instruction, 1900–1930. "Perdóname míster, me puede decir donde queda la bus station?" E»S

mistificación (mees‑tee‑fee‑KA‑zion), n., f., mystification. E»S

mistificar (mees‑tee‑fee‑KAR), v., to mystify. "El se mistifica cada vez que escucha la voz de Janis Joplin." E»S

mistihueso (mees‑tee‑WE‑so), n., f., Smith and Wesson pistol. First recorded by Pedro Henríquez Ureña.

mitin (MEE‑teen), n., m., 1. business meeting. 2. public act in which orators give speeches. E»S

mix (MEEX), n., m., juxtaposition. "Carlos es un mix de Puerto Rican y de African‑American."

mixtar (MEEX‑tar), v., to mix. E»S Also MIXTEAR.

mixtear (MEEX‑tar), v., to mix. E»S Also MIXTAR.

moba (MO‑ba), n., m/f., mob. "Tony Soprano pertenece a la moba." E»S

móbil (MO‑beel), n., m., mobile phone. "I will call you from el móbil." E»S

mobilhom (mo‑beel‑HOWM), m., f., mobile home. E»S

mobilidad (mo‑bee‑lee‑DAD), n., f., social mobility. E»S Also MORBILIDAD.

mocasín (mo‑ka‑SEEN), n., m., moccasin.

módem (MO‑dem), n., m., modular connector. {CS} E»S

mofin (MO‑feen), n., m., muffin. Sp. *panque* and *panqueque*. "Todas las mañanas we eat mofines." E»S

mofle (MO‑fleh), n., m., muffler. E»S

mog (MOG), n., f., large coffee cup. "Dame el capuccino en una mog." E»S

mohair (mo·ha·EER), n., m., material made with Angora goat hair.

moisteraizer (moy·ste·RAI·zer), v., moisturizer. {CA} E»S

moisterizar (moy·ste·ree·ZAR), v., to moisturize. "Ella se moisteriza the face." {CA}

mojo (MO·kho), n., m., little kid. "The place was full of 12-year-old mojos." {Ch} Also SHAMAC.

mol (MOLL), n., m., mall. "Las adolescentes are having fun en el mol." E»S

mona (MOH·na), n., f., 1. beauty. "The girl está mona!" 2. tops, as in children's game. {Ch}

moni (MO·nee), n., m., money. E»S Also MONIS.

moniorder (mo·nee·OR·der), n., m., money order. E»S

monis (MO·nees), n., m., money. E»S Also MONI.

monitor (mo·nee·TOR) n., m., TV monitor.

monte (MON·te), exp., sure thing. The purchase of land es un monte." The terms dates from the 19th century. {SW} Also MONTI.

monti (MON·tee), exp., 'sure thing.' The purchase of land es un monti." The terms dates from the 19th century. {SW} Also MONTE.

mopa (MO·pa), n., f., mop. "¿Donde está la mopa to clean the mess on the floor?" E»S Also MOPEADOR.

mopeador (mo·pea·DOR), n., m., mop. "¿Donde está el mopeador to clean this mess on the floor?" E»S Also MOPA.

mopear (mo·PEHAR), v., 1. to map. "Pedro mopea la ruta de Nueva York a Boston." 2. to mop. "Make sure to mopear the kitchen." E»S Also MAPEAR, MAPIAR and MOPIAR.

mopiar (mo‑PEHAR), v., 1. to map. "Pedro mopia la ruta de Nueva York a Boston." 2. to mop. "Make sure to mopiar the kitchen." Also MAPEAR, MAPIAR and MOPEAR.

mora (MO‑ra), n., f., girl. "That mora knows how to dance." {LA}

morami (mo‑RAH‑mee), n., m., mortar mix. {CA}

mórbido (MOR‑bee‑do), adj., morbid. E»S

morbilidad (mor‑bee‑lee‑DAD), n., f., mobility. "La familia de Diego tiene morbilidad social. Ya compraron una casa nueva." E»S Also MOBILIDAD.

morro (MO‑rro), n., m., 1. little brother. 2. boy. "Hey morro, get the hell out of here."

mosquito (mos‑KEE‑to), n., m., bothersome person. "Pedro es el mos‑quito de la clase." {Ch}

motel (mo‑TEL), n., m., highway hotel. "Ayer dormimos en un motel."

motocrós (mo‑to‑CROS), n., f., athletic activity in which motorcycles are used for climbing steep terrain. E»S

mous (MAUS), n., m., computer mouse. "Mueve el mous para abrir la window." {CS} E»S Also MAUS and MOUSE.

mouse (MAUS), n., m., computer mouse. "Mueve el mouse para abrir la window." {CS} E»S Also MAUS and MOUS.

moverse (mo‑VER‑se), v., to move from one geographic location to an‑other. Sp. *moverse* geographically is *mudarse.* E»S

mula (MOO‑la), n., f., money. "Do you need mula to go to the club?" {Ch}

multifario (mool‑tee‑FA‑rio), adj., multifarious. "Su padre tiene talentos multifarios." E»S

multimedia (mool‑tee‑ME‑dia), adj., multimedia. E»S

multipropósito (mool·te·pro·PO·see·to), n., m., involved in multiple purposes. "This mechanical tool es multipréposito." E»S

multitasco (mool·tee·TAS·ko), n., m., multitasking. E»S

muñequita (moo·ye·KEE·ta), n., m., beer. After the St. Pauli Girl brand name. {CA}

music (MOO·sik), n., f., American·style popular music. "A ellos les gusta la music de los cincuenta." E»S

music hall (moo·seek·HALL), n., m., variety show. "Fuimos a ver el music hall en el Radio City." E»S

muzac (MOO·zac), n., f., kitschy music. "En la oficina del doctor siem·pre tocan muzac." E»S

N

nacho (NA‑tcho), n., m., tortilla chip. "She serves nachos with cheese as appetizer."

naco (NA‑koh), adj., and n., m/f., pleb, person of the lower‑class. "The nacos live en the hood." {Ch}

nai (NAY), adj., nice. "Este suéter está nai." {NY}

nailón (nay‑LON), n., m., 1. plastic bag. "Mítele la cazuela de potaje a tu abuela en un nailón pa' que no se vire en el carro." 2. panty‑hose. "Ella se ve bien con los nailón." E»S

naitclub (NAIT‑club), n., m., night club. E»S

narco (NAR‑koh), n., m/f., drug trafficker.

nativo (na‑TEE‑vo), n., m/f., native. "She is nativa de Puerto Rico." Sp. *oriundo*. E»S

nebulizar (ne‑boo‑lee‑ZAR), v., to spray paint.

nectarina (nec‑ta‑REE‑na), n., f., nectarine. From Eng. and Fr. *nectarine*.

nerlandés (ner·lan·DEZ), n., m/f., negative individual. "Juan siempre busca errores. Es nerlandés." From Eng. *nay-sayer.* {Ch, SW} E»S

nefario (ne·FA·rioh), adv., nefarious. "Esta situación es nefaria." E»S

Neorican (ne·oh·REE·kan), n., m/f., Puerto Rican person in the mainland. Sp. *Nuevo Riqueño.* "El dueño del almacén de comida latina es un Neorican." Also NUYORICAN and NUYORRIQUEÑO.

nerdear (ner·DEHAR), v., to behave like a nerd. "Are you still nerdando, even though it is nice outside?" Also NERDIAR. E»S

nerdiar (ner·DEEAR), v., to behave like a nerd. "Are you still nerdiando, even though it is nice outside?" Also NERDEAR. E»S

nerdio (ner·DIOH), n., m/f., nerd. "Swarthmore College es una factoría de nerdios." E»S

nesteado (nes·TEAH·do), adv., m/f., 1. stuck. 2. grounded. "Guillermina tuvo un problema con her car y está nosteada." E»S

net (NET), n., f., internet. The preferred term is *la Web.* {CS}

neta (NE·ta), n., f., truth. "Jaime said the neta about the incident." {Ch}

ninja (NEEN·jah), n., m/f., ninja.

niquel (NEE·kel), n., m., nickel. E»S Also NIQUER.

niquer (NEE·ker), n., m., nickel. E»S Also NIQUEL.

nítido (NEE·tee·doh), adj., to one's liking. "Esa ranfla está nítida!" {LA, NY}

niueig (noo·WAIG), n., m., New Age. "Ese estilo es niueig." E»S

niuluc (NOO·looc), n., m., new look. E»S

niusleter (NOOS·le·ter), n., f., newsletter. E»S

niuwaif (NOO·waif), n., f., New Wave. E»S

nocaut (no·KAUT), n., m., knockout. Boxing term. "¿Viste el nocaút que recibió el boxeador?" E»S

nocautear (no‑kau‑TEAR), v., to knock out a fighter. E»S Also NO‑QUEAR.

nocdaun (NOK‑dawn), n., m., knockdown. Boxing term. E»S

nominación (no‑mee‑na‑ZION) n., f., nomination. "La película obtuvo una nominación para el Oscar." E»S

nominado (no‑mee‑NA‑do), adj., m/f., nominated. "Ese filme está nominado para two Oscars." E»S

nominar (no‑mee‑NAR), v., to nominate. E»S

noquear (no‑KEAR), v., to knock out. Boxing term. E»S Also NO‑CAUTEAR.

Norgué (nor‑GUEH), n., northwest section of Miami. E»S

norsa (NOR‑sa), n., m/f., nurse. E»S

Norte (NOR‑teh), n., m., name for the United States. Usually known as *El Norte.*

nouhau (NOU‑haw) n., m., know‑how. E»S

noutbuk (NOUT‑book), n., m., portable computer. {CS} E»S

nulificar (noo‑lee‑fee‑KAR), v., to nullify. "The check quedó nulificado." E»S

Nuyersi (NOO‑yer‑see), n., New Jersey. {CA}

Nuyol (NOO‑yol), n., New York. {NY}

Nuyorican (nu‑yo‑REE‑kan), n., m/f., Puerto Rican person in the mainland. The term is also spelled Nuyorrican. Sp. *Nuevo Riqueño.* Also NEORICAN and NUYORRIQUEÑO.

Nuyorquino (noo‑yor‑KEE‑no), n., m/f., New Yorker.

Nuyorriqueño (nu‑yo‑ree‑KE‑nyo), n., m/f., Puerto Rican person in the mainland. Sp. *Nuevo Riqueño.* Also NEORICAN and NUYORICAN.

Ñ

ñato (NYA-to), n., m., guy.

ñero (NYEH-roh), n., m., a Mexican-American male. {SW} Also POCHO.

ñango (NYAN-goh), adj., m/f., scrawny.

ñudo (NYOO-doh), n., m., knot.

O

oben (OH-ven), n., m., oven. {Ch} E»S

obligatoriedad (oh-blee-ga-to-reih-DAD), n., f., forcefulness. E»S

obliterar (oh-blee-te-RAR), v., to obliterate. "César obliteró the important information." E»S

obnoxio (ob-KNO-xioh), n., m/f., obnoxious. E»S

obsolescente (ob-soh-le-ZEN-te), n., m/f., obsolete. E»S

occurrancia (oh-koo-REHN-cia), n., f., occurrence. E»S Also OCURENCIA.

occurrir (oh-koo-REER), tv., to occur, to take place. "El accidente OCCURRIÓ en Fifth Avenue." E»S

ocupacional (o-KOO-pa-tio-nal), n., m., occupational therapy. E»S

ocurrencia (oh-koo-REHN-cia), n., f., occurrence. E»S Also OCCURANCIA.

of (OVH), adj., m., with electric power interrupted. "El switch está of." E»S

of-de-recor (ohv-the-REH-cord), exp., information transmitted privately. E»S

ofensa (oh-FEN-zah), n., f., offense. E»S

offshore (OFF-shorh), n., m., 1. petroleum rigs. 2. investment company in foreign soil. E»S

oficial (oh-FEE-tial), n., m/f., police. E»S

ofisdei (OFEES-dey), n., m., office day. E»S

ofiz (OH-fiz), n., f., office. E»S

ofline (of-LAYN), exp., off line. {CS} E»S

of-moder-de-cais (ohv-mooh-der-the-kaiz), exp., out of hand activity. Literal translation of Sp. *de madre el caso*. "Pepe está al borde del bankrup. Ay, of-moder-de-cais!" {CA} E»S

ofsaid (OVH-sayd), n., m., outside of the game. E»S

ofsé (ohv-SEH), n., m., 1. printing press. 2. unhappy. "Ella esta ofsé." E»S

okai (oh-KEY), exp., OK. from Eng. Also OQUE. E»S

olrait (all-RAID), exp., all right. E»S

on (ohn), adj., with on-going electrical power. "El switch está on." E»S

onda (ON-dah), exp., fashion. E»S

ondebich (on-deh-BEECH), exp., without goal. E»S

on-de-cuidado (OHN-de-kuee-DA-doo), exp., on the lookout. {Ch}

ondergraun (on-THER-groun), n., m., 1. avant-garde art. 2. subway system. E»S

onlain (on-LAYN), exp., online, connected. {CS} E»S

onrasonable (un-ree-zo-nah-BLE), n., m., unreasonable. E»S

open (OH‑pen), n., m., non‑restrictive sport competition. Used to refer to the U.S. Tennis Tournament. E»S

oqué (oh‑KE), exp., OK. from Eng. Also OKAI.

órale (OH‑rah‑le), exp. of agreement. "Orale vato, of course. I'll go with you." {Ch}

orejano (oh‑re‑KHA‑noh), n., m., unmarked, unbranded cow. {SW}

orsay (OOHR‑say), n., m., outside boundaries. E»S

othercide (oh‑THER‑sayd), n., m., the elimination of people or attrib‑ utes different to us or ours. "Lord Jeffrey Amherst organizó un othercide of Native Americans." E»S

out (AU‑th), exp., 1. out of fashion. 2. obsolete. 2. given a last chance. Athletic term. "Esa pelota está out." Also AUT.

outin (OUH‑teen), n., m., making public a private fact. E»S

output (ouht‑POOT), n., m., outgoing data. {CS} E»S

outsider (out‑ZAY‑der), n., m., unknown, foreigner. E»S

outsorcin (OUT‑sohr‑seen), n., m., massive subcontracting. {CS} E»S

over (OH‑ver), exp., 1. on top. "Nuestra casa está over la montaña." 2. finished. "Ella está over su amor con Valentín."

overbuc (ooh‑VER‑buck), n., m., outnumbered. "The flight está over‑ buc." E»S

overbukear (ooh‑ver‑BOO‑kear), v., to overbook. "Marta overbukeó su horario." E»S Also OVERBUKIAR, OVERBUQUEAR and OVERBUQUIAR.

overbukiar (ooh‑ver‑BOO‑kear), v., to overbook. "Marta overbukió su horario." E»S Also OVERBUKEAR, OVERBUQUEAR and OVERBUQUIAR.

overbuquear (ooh‑ver‑BOO‑kear), v., to overbook. "Marta overbuqueó su horario." E»S Also OVERBUKEAR, OVERBUKIAR and OVERBOQUIAR.

overbuquiar (ooh-ver-BOO-kear), v., to overbook. "Marta overbuquió su horario." E»S Also OVERBUKEAR, OVERBUKIAR and OVERBUQUEAR.

overflo (oh-ver-FLOH), n., m., overflow. {CS} E»S

overmirar (ooh-ver-MEE-rar), v., to overview.

overo (oh-VEER-oh), n., m., a pinto horse found in Argentina. E»S

overol (oh-VER-all), n., m., p., heavy-duty clothes. Eng. *overall.* E»S

ozomatli (OH-zo-mah-TLI), n., f., Chicano music with the social consciousness of the 1970s, also known as World Beat music. {Ch}

P

pac (PAK), n., m., package. E»S Also PAK.

pachuco (pah-TCHOO-koh), n., m/f., 1. a Chicano person in East Los
Angeles. 2. A Chicano youth in Los Angeles or another southern US city, caught between Mexican and US cultures. The expression was used in the forties in California. {Ch, SW}

pachwork (PATCH-work), n., m., composite work. E»S

padle (PAH-del), n., m., athletic activity similar to tennis but played
with smaller wooden raquets. E»S

padloc (PAD-lok), n., m., enclosure for horses before a race.

page (PAIG), n., f., internet page. E»S Also PEIG.

pagear (pa-GEAR), v., to page. {CS} E»S Also PEIGEAR

página-de-casa (PA-gee-na de KA-sa), n., f., home internet page. {CS}

pak (PAK), n., m., package. E»S Also PAC.

palacial (pa-la-ZIAL), adj., palatial. "That house es palacial." E»S

palé (paah-LEEH), n., m., pallet.

palmtop (PALM-top), n., m., portable computer. {CS} E»S

palomilla (pa-loh-MEE-ya), n., f., 1. an all milk-white or cream-colored horse. "Rancho San Isidro has a beautiful three-year-old palomilla for sale." 2. group of friends. "Vino la palomilla. The group arrived." {SW}

Pambich (PAM-beech), n., Palm Beach. {CA} E»S

pamper (PAM-pehr), n., m., diaper. After Pamper brand name. E»S Also PANPER.

panfletear (pan-fle-TEAR), v., to be a pamphleteer. E»S Also PANFLETIAR.

panfletiar (pan-fle-TEAR), v., to be a pamphleteer. E»S Also PANFLETEAR.

panfletista (pan-fle-TEES-ta), n., m/f., pamphleteer. E»S

panfleto (PAN-phe-tho), n., m., pamphlet. E»S

panikear (pah-nee-KEAR) v., to panic. "He didn't know what to do y se panikeó." E»S Also PANIKIAR.

panikiar (pah-nee-KEAR) v., to panic. "He didn't know what to do y se panikió." E»S Also PANIKEAR.

pankeis (pahn-KAYS), n., m., breakfast pancakes. "I like nothing better para el desayuno que los pankeis." E»S

panper (PAN-pehr), n., m., diaper. After brand name. E»S Also PAMPER.

pans (PAHN-tz), n., m., warm-up pants. Always used in the plural. E»S

panty (PAHN-tee), n., m., nylon leotards. E»S

paqueguín (pa-keeh-GIN), n., m., marketing technique. "Coca-Cola usa un paqueguín muy ingenioso to sell its products." E»S

paqueque (pa-KE-keh), n., m., pound cake. E»S

paquete (pah KEH teh), n., m., stack. {Ch}

parcadero (par-kea-DE-roh), n., m., parking garage. E»S Also PAR-QUADERO and APARCADERO.

parchisi (par-CHEE-see), n., m., parcheesi. E»S

pari (PA-ree), n., f., a party. "La pari de Rosa is tomorrow."

parisear (pa-ree-SEAR), v., 1. to party. 2. to hang around parties.

parisero (pa-ree-ZE-roh), n., m/f., party-goer. E»S "Juan es un parisero en el wikén."

parkear (par-KEAR), v., to park. "No había sitio para parkear at the marqueta." E»S Also APARCAR, PARKIAR, PARQUEAR and PARQUIAR.

parkiar (par-KEAR), v., to park. "No había sitio para parkiar at the mar-queta." E»S Also APARCAR, PARKEAR, PARQUEAR and PARQUIAR.

parkin (PAR-keen), n., m., the act of parking the automobile. E»S

parqueadero (par-kea-DE-roh), n., m., parking garage. E»S Also PAR-CADERO and APARCADERO.

parquear (par-KEAR), v., to park. "No había sitio para parquear at the marqueta." E»S Also APARCAR, PARKEAR, PARKIAR and PARQUIAR.

parqueo (par-KE-o), n., m., 1. parking zone. 2. act of parking an auto-mobile.

parquiar (par-KEAR), v., to park. "No había sitio para parquiar at the marqueta." E»S Also APARCAR, PARKEAR, PARKIAR and PARQUEAR.

parsec (PAR-sec), n., m., per second. E»S

partain (par-TAIN), n., m., part-time job. {CA}

partición (part-tee-ZION), n., m., partition. "Hubo una partición después de la Guerra México-Norteamericana." E»S

pasage (pa-ZA-ge), n., m., route. "El no encontró el pasage to Arizona." Sp. *pasaje*. E»S

pasinshot (pa-seen-SHOT), n., m., quick shot. Tennis term. E»S

paso (PAH-soh), n., m., mountain pass, fjord. {SW}

pastar (pas-TAR), v., to paste. In Sp. *pastar* means to graze. E»S

pastear (pas-TEAR), v., to paste. {CS} E»S Also PASTAR.

patio (PA-tee-oh), n., m., open space inside the house. The term dates from 1828. S»E

patrol (pa-TROL), n., m., border patroling police. E»S

patronista (pa-tro-NEES-ta), n., m/f., patron of the arts and humanities. E»S

patronizar (pa-tro-nee-CZAR) v., to patronize. "Martín patroniza el restaurante mexicano." E»S

pauperismo (pau-pe-REES-mo), n., m., pauperism.

peccadillo (pe-ka-DEE-yo), n., m., venial fault. The term dates from 1591. S»E

pedigri (pe-dee-GREE), n., m., distinguished lineage. E»S

pedo (PE-do), n., m., 1. trouble. 2. drunken state. "Qué pedo I got myself into last night!" {Mex}

pégamelas (PEH-ga-me-las), exp., Give me five!

pei (PEY), n., m., payment.

peig (PEIDG), n., f., internet page. E»S Also PAGE

peigear (pei-DGEAR), v., to page. E»S Also PAGEAR.

peigviu (PEIDG-vee-u), n., m., Internet page on display. E»S {CS}

peiperback (pei‑per‑BAK), n., m., paperback book. "He read la novela en un peperback." E»S

pelado (pe‑LA‑do), n., m/f., ill‑bred person. {Ch, Mex}

penal (pe‑NAL), n., m., penalty kick. Soccer term. E»S Also PENALTI.

penalizar (pe‑na‑lee‑ZAR), v., to penalize. E»S

penalti (PE‑nal‑tee), n., m., penalty kick. Soccer term. E»S Also PENAL.

penbook (PEN‑book), n., m., portable computer with special type of pencil with which one writes on the screen. {CS}

penpal (pen‑PAL), n., m/f., correspondent. "Mariana tiene un penpal en Pakistán." E»S

péo (PEH‑oh), n., m., P.O. Box. E»S

peon (PEE‑on) n., m., common worker. Sp. *peón*. E»S

pepermín (PE‑per‑meen), n., m., pepper‑mint, mint liquor which is a mix of alcohol, mint, and sugared water. E»S

performance (per‑FOR‑manz), n., f., 1. act. "Guillermo está en un per‑formance en Broadway." 2. pattern. "El performance de ese producto is as expected." 3. business deal. "La compañía hizo un performance y compré más tools." E»S

performeador (per‑for‑mea‑DOR), n., m/f., performer. E»S Also PER‑FORMIADOR.

performear (per‑for‑MEHAR) v., to perform. "Tenemos que restar nuestras voces esta noche porque performeamos mañana." E»S Also PERFORMIAR.

performiador (per‑for‑mea‑DOR), n., m/f., performer. E»S Also PER‑FORMEADOR.

performiar (per‑for‑MEHAR) v., to perform. "Tenemos que restar nuestras voces esta noche porque performiamos mañana." E»S Also PERFORMEAR.

perfuctorio (per‑fook‑TO‑rio), adj., perfunctory. E»S

periodical (pe‑rio‑DEE‑kjal), n., m., newspaper, magazine. "Leímos la noticia en el periodical." E»S

permision (per‑mee‑ZION), n., m., permission. In Sp. the appropriate term is *permiso.* E»S

perro caliente (PE‑rro ka‑LIEN‑te), n., m., hot dog. E»S

persura (per‑ZOO‑ra), n., f., pursuit. "Ella decidió hacer una persura in order to find la verdad." E»S

peso (PEH‑soh) n., m., used in place of *dólar* to mean dollar. S»E

photofinish (fo‑to‑FEE‑nish), exp., 1. competitive ending to a race. 2. quality of competition. E»S

picadillo (pee‑ka‑DEE‑yo), n., m., 1. hors d'oeuvre. 2. mashed meat. "Comimos picadillo para el lunche." S»E

picap (pee‑KAP), n., m., record player. (Chi} E»S

pícaro (PEE‑ka‑roh), n., m., someone on the make, prowling for easy money. The term dates from the 18th century. In Sp. *pícaro* means a rogue. Eng. *picaroon* and Sp. *picarón*

pichear (pi‑TCHIAR), v., to pitch. "¿Quién va a pichear in tomorrow's game?"

pícher (PEE‑tcher), n., m., pitcher. Baseball term. "Ni ketcha, ni pitcha, ni deja batear." E»S

pickear (PEE‑kyar), v., to select, to make a choice. "Pickea uno vato, and let's leave." E»S Also PIKIAR.

pickup (PEE‑kup), n., f., 1. pick-up truck. 2. record player. E»S

picle (PEE‑kle), n., m., sour pickle. {NY} E»S

pícnic (PIK-nik), n., m., outdoor meal. "Yesterday la familia se fue de picnic." E»S

picop (PEE-cop), n., f., pick-up vehicle. E»S

pijama (pee-KHA-ma), n., f., pajamas. E»S

pílin (PEE-lean), n., m., skin treatment through which dead skin is removed. E»S

pimento (pee-MEN-toh), n., m., 1. dried aromatic berries. 2. type of chile. The term dates from 1690. S»E

pimpo (PEEM-poh) n., m., pimp. "Hollywood is crawling with pimpos." E»S

pin (PEEN), n., m., clothing decoration. "Ella usa pins con mensajes antibélicos." E»S

pinta (PEEN-tah), n., f., prison. "Mi primo está en la pinta but it isn't his fault." {LA}

pinto (PEEN-toh) n., m., spotted horse. Originally from Sp. *pintar*, to paint. "Is that pinto over there female or male?" S»E

pinup (peen-UP) n., f., female beauty. E»S

pínut (PEE-nut), n., m., peanut. "Me gusta la mantequilla de pínut." Sp. *maní* and *cacahuate*. E»S

piquinini (pee-kee-NEE-nee), n., m., little boy. "El piquinini de mi hermana is already three years old." E»S

pírcin (PEER-seen), n., m., piercing, the practice of making perforations to wear pendants in any part of the body. E»S

placa (PLAH-kah) n., f., cop, police officer. "Careful, bro! There is a placa behind us." {Ch}

placete (pla-ZE-te), n., m., place, site. "Nos conocimos en un placete de San Antonio." E»S

planin (PLA-neen), n., m., elaboration of detailed diagrams in which different work variables are written. E»S

plano (PLA-no), adj., m/f., 1. plan. "Hagámos un plano para resolver el problema." 2. n., m., geographical plain. "Beyond those mountains está el plano." E»S

planta (PLAN-tah), n., f., architectural complex. In Sp. *planta* means vegetable. E»S

plasearse (pla-SEAR-se), v., to please oneself. "Ella se plasea con ropa muy fina." E»S

plastero (plas-TE-roh), n., m., plasterer. E»S

plástico (PLAS-tee-ko), n., m., credit card. "Usa el plástico para el purchase." E»S Also TARJETA.

plate (PLAIT), n., m., home plate. Baseball term. E»S Also PLEITT.

platitud (pla-tee-TOOD), n., f., irrelevant comment. "Lo que dijo ella fue una platitud." E»S

play (PLEY), n., m., in technology, the button designated as *play*, which activates the mechanism. "Apachurra el play para oír la música." E»S Also PLEI.

playback (pley-BAK), n., m., the reactivation of previously recorded material. E»S

playboy (pley-BOY), n., m., gigolo. "Javier es un playboy." S»E

playof (pley-OF), n., m., final stages of a competition, play-off. Sports term. E»S

plaza (PLA-za), n., f., marketplace, public place. The term dates from 1683. S»E

pleasura (ple-a-SOO-ra), n., f., pleasure. "Don Quixote felt pleasura en su casa." E»S

plegiar (ple-GEAR), v., to pledge. E»S

plei (PLEY), n., m., in technology, the button designated as *play*, which activates the mechanism. "Apachurra el plei para oír la música." E»S Also PLAY.

pleit (PLAIT), n., m., home plate. Baseball term. E»S Also PLATE.

ploguear (plo-GEAR), v., to plug in. E»S Also PLOGUIAR.

ploguer (PLO-guer), n., m., connecting device. E»S

ploguiar (plo-GEAR), v., to plug in. E»S Also PLOGUEAR.

plomcaik (PLUM-kaik), n., m., plum cake.

plote (PLO-the), n., m., plot. "I don't understand el plote del cuento."

plotter (PLO-ter), n., m., data-processing machine for tracing complex graphics.

pob (POB), n., m., pub, bar.

poblicrelación (PO-blik re-LAY-zion), n., m., effort to reach an audience, public relations. "La compañía hace poblicrelación para tener más clientes." E»S

poca (PO-ka), n., m., poker. Also POQUER. E»S

pocas-labras (PO-kas-LAH-brahss), exp., few words. "He is a man of pocas-labras." Also POCO. S»E

pochismo (poh-TCHEES-mo), n., m., ideology behind the Pocho attitude toward life. Also used to describe Spanglish-style attitude, e.g., mixing Spanish and English. {Mex, SW}

pochista (poh-TCHEES-ta), n., m/f., a person who consciously mixes Spanish and English but happily lives between two worlds. {SW}

pocho (POH-tcho), n., m/f., 1. a Mexican person assimilating to the culture of the United States. 2. a person from the United States of Mexican descent, but who grew up in the States. The term has a negative connotation of being materialistic. Often used by Mexicans offensively about Chicanos. "Hey pocho, go back to California." {Mex, SW}

poco (PO·koh) adj., little. "In the land of poco tiempo." S»E

poco·malo (poh·coh MAH·loh), exp., not too bad. "This restaurant es poco·malo." S»E

pointear (poyn·TEAR), v., to point to, to point at. E»S Also POIN·TIAR.

pointer (POYN·ter), n., m., a type of dog.c

pointiar (poyn·TEAR), v., to point to, to point at. E»S Also POIN·TEAR.

poliéster (po·LYES·ter), n., m., synthetic fabric, polyester. E»S Also POLIESTIRENO and POLYESTER.

poliestireno (po·lyes·Tee·RE·no), n., m., synthetic fabric, polyester. E»S Also POLIESTER and POLYESTER.

polís (po·LEES), n., f., police. {NY} E»S

polishear (po·lee·SHEAR), v., to polish. E»S Also POLISHIAR.

polishiar (po·lee·SHEAR), v., to polish. E»S Also POLISHEAR.

político (po·LEE·tee·ko) n., m., 1. immoral and ambitious politician. 2. any politician. Used in the United States since the early 20th century. S»E

polposición (pol·po·ZEE·zion), n., f., in a motorcycle or automobile race, the take·off of the first competitor.

polyester (po·LYES·ter), n., m., synthetic fabric, polyester. E»S Also POLIESTER and POLIESTIRENO.

pomelo (po·ME·lo), n., m., pomelo, a type of fruit. S»E

pompa (POM·pah), n., f., pump. "Usa la pompa del agua, por favor." E»S

ponch (PONCH), n., m., hit. Used as a boxing term.

ponchar (pon·TCHAR), v., to punch. E»S Also PONCHEAR.

ponchazo (pon·TCHA·zo), n., m., huge punch. E»S

ponchear (pon·TCHEAR), v., to punch. E»S Also PONCHAR.

poncho (PON·tcho), n., m., shawl worn as an outer garment. Also SERAPE. S»E

poner énfasis (po·NER EN·pha·sees), v., to place emphasis. E»S

poni (POH·nee), n., m., pony horse.

pop (POP), adj., popular. The term is used in music to describe a type of song appreciated by the masses. E»S

poper (PO·per), n., m., type of synthetic and hallucinogenic drug that is inhaled.

populación (po·poo·la·ZION), n., f., population. Sp. *población.* E»S

poquebú (po·ke·BOO), n., m., pocket book. {NY} E»S

póquer (PO·ker), n., m., poker, card game. E»S Also POCA.

pora (POH·ra), n., m., porter. E»S

porche (POR·tche), n., m., patio. E»S

portable (por·TA·ble), adj., portable. "He has una TV portable." E»S

portorriquéno (por·to·ree·KEE·nyo), n., m/f., person from Puerto Rico. Also BORICUA, BORINCANO, BORINQUEÑO and PUERTORRIQUEÑO.

portfolio (PORT·fo·lio), n., m., display of images or information. In Sp. *portafolio* means *business suitcase.*

posada (po·SA·da) n., f., inn or roadhouse. "Mi mother stays in a posada when she travels." S»E

posbalance (pos·BA·lanz), n., m., period immediately following the re-alization of an annual balance. E»S

posesivo (po·se·SEE·vo), adj., possessive. E»S

posición (po·see·ZION), n., f., position of a job in the corporate or in-

stitutional scale. "Juan tiene una posicion buena en la compañía." E»S

postear (pos-TEAR), v., to post. E»S

posteo (pos-THE-oh), n., m., posting. E»S

póster (POS-ter), n., m., large sign. E»S

postit (POS-teet), n., m., small sheet of sticky paper, yellow note-paper.
After stationery brand. E»S

potro (PO-tro), n., m., colt, filly. "That feisty little potro will never be
a good workhorse."

pouzol (POW-zol), n., m., puzzle. Sp. *rompecabezas.* "El niño es un
genio. Hizo el pouzol en ten minutes." E»S Also POZEL and
PUZLE.

pozel (PO-zel), n., m., combination game, puzzle. Sp. *rompecabezas.* E»S
Also POUSOL and PUZLE.

pragmatismo (prag-ma-TEEZ-mo), n., m., Americanism.

pragmatista (prag-ma-TEES-ta), n., m/f., person who professes the phi-
losophy of pragmatism.

praisear (pray-SEAR), v., to praise. E»S Also PRAISIAR.

praisiar (pray-SEAR), v., to praise. E»S Also PRAISEAR.

preciado (pre-ZIA-doh), adj., precious. "Esta comida es preciada." Sp.
apreciado. E»S

predatorio (pre-da-TO-rio), adj., predatory. E»S

pregneada (pre-GNA-da), adj., f., pregnant. "Don't be a fool, está
pregneada." {Ch} E»S

prejuiciado (pre-khuee-ZIA-do), n., m/f., in a stage prior to the legal or
criminal trial Sp. *perjudicia* means prejudiced. E»S

prejuicio (pre-KHUEE-zioh), n., m., hearing prior to the legal or criminal trial. In Sp. *perjuicio* means prejudice, bias E»S

premier (PRE-mee-er), n., m., chief of state, prime minister. "Tony Blair es el premier de Inglaterra." E»S

presente (pre-SEN-te), n., m., gift. "En mi cumpleaños mi sister me dio un presente." Sp. *presente* means present, as in to be present. Also, gift is *regalo.* E»S

presidio (pre-SEE-dyo), n., m., garrison town. The term dates from 1808. S»E

presin (PRE-seen), n., m., the act of pressing. "El hace un presin en el botón del elevador." E»S

presura (pre-SOO-ra), n., f., pressure. "Siento mucha presura en casa." E»S

presurizar (pre-zoo-ree-CZAR), v., to pressurize, referring to an enclosed space. 2. to maintain the atmospheric pressure adequate for humans.

prevaricación (pre-ba-ree-ka-ZION), n., f., act of prevaricating. E»S

prevaricador (pre-ba-ree-ka-DOR), n., m/f., person who prevaricates.

prevaricar (pre-ba-ree-KAR), v., to prevaricate. E»S

previsar (pre-vee-SAR), v., to preview. "Las distribuidoras de Hollywood previsan películas para despertar interés in the audience." E»S

previsión (pre-vee-ZION), n., f., preview. "La compañía dará una previsión del nuevo producto esta mañana." E»S

prime-rate (PRAYM-rait), n., f., preferential interest rate. E»S

prime-time (PRAIM-taym), n., m., maximum-audience period on TV. "Jorge tiene su programa en prime-time." E»S

prinada (pree-NAH-dah), n., f., female swindler. The term dates from 1620. Origins possibly from Sp. *preñada*, which means pregnant. "She is a cunning prinada that swiped my purse and I didn't even notice it."

prinador (pree·na·DOR), n., m., pimp. Origin possibly from Sp. *preñador.*

principal (preen·ZEE·pal), n., m/f., school principal. In Sp. *principal* means main, leading. E»S

printear (PRIN·tear), v., to print. E»S Also PRINTIAR.

printer (PREEN·ter), n., f., printing devise. "Voy a imprimir el documento en la printer."

printiar (PRIN·tear), v., to print. E»S Also PRINTEAR.

procastinador (pro·cas·tee·na·DOR), n., m/f., procastinator.

procastinar (pro·cas·TEE·nar), v., to procrastinator. E»S

product manager (PRO·duct MA·na·ger), n., m/f., person in charge of organizing a merchandizing effort. "El es el product manager de la empresa." E»S

profit-teiken (PRO·feet TEI·ken), n., m., revenue taken away in financial transaction.

progresivo (pro·gre·SEE·vo), adj., politically progressive. Also used as noun. "El candidato es un progresivo." E»S

proliferear (pro·lee·fe·REAR), to multiply. E»S Also PROLIFERAR and PROLIFERIAR.

proliferar (pro·lee·fe·REAR), to multiply. E»S Also PROLIFEREAR and PROLIFERIAR.

prolisferiar (pro·lee·fe·REAR), to multiply. E»S Also PROLIFEREAR and PROLIFERAR.

promisa (pro·MEE·sa), n., f., promise. Sp. *promesa.* E»S

promoción (pro·mo·ZION), n., f., promotion. "Le dieron una promoción." In Sp. *promoción* means *advertising.* E»S

promtear (prom·TEAR), v., to prompt. E»S

promteador (prom·TEA·dor), n., m/f., TelePrompter. E»S

prontear (pron·TEHAR), v., to accelerate.

pronto (PRON‑to), adv., 1. at once, quickly. 2. immediately. "Come here, pronto!" S»E

propela (pro‑PE‑la), n., f., propellor. E»S

prospecto (pro‑PEK‑to), n., m., candidate, prospect. E»S

prototipear (pro‑to‑pee‑PEHAR), v., to prototype. E»S

providear (pro‑vee‑DEHAR), v., to provide. E»S

providor (pro‑vee‑DOR), n., m/f., provider. Sp. *provedor.* E»S

psicodélico (psy‑ko‑DE‑lee‑ko), adj., psychedelic. E»S

psicokiller (psy‑ko‑KEE‑ler), n., m./f., a psychotic killer. E»S

publicador (poo‑blee‑ka‑DOR), n., m., publisher. E»S Also PUB‑LISHER

publisher (poo‑blee‑SHER), n., m/f., 1. publisher. 2. editor. Also PUBLICADOR.

puchar (pun‑TCHAR), v., to push. E»S Also PUCHEAR.

puchear (pun‑TCHEAR), v., to push. E»S Also PUCHAR.

pudín (poo‑DEEN), n., m., desert pudding made of egg. Also PUDINGA.

pudinga (poo‑DEEN‑ga), n., m., desert pudding made of egg. Also PUDIN.

pueblo (PUE‑bloh), n., m., Spanish village, Indian village. The term dates from 1818. S»E

puertorriqueño (puer‑to‑ree‑KE‑nyo), n., m/f., person from Puerto Rico. Also BORICUA, BORINCANO, BORINQUEÑO and PORTORRIQUEÑO.

pul (POOL), n., m., 1. grouping of item, people, or societies. 2. swimming pool. 3. pool, billiards. 4. influence. "Juanito tenía un pul en el INS y por eso consiguió el grinca." E»S Also POOL.

pulear (poo‑LEAR), v., to pull. E»S

pullman (POOL-man), n., m., railroad sleeping car. E»S

púlover (POOL-oh-ver), n., m., a sweat-shirt. "It's cold outside, vas a nececitar un púlover." {Arg} E»S Also PULLOVER.

pulsar (pool-ZAR), n., f., a neutron star emitting radioelectronic impulses. "Vimos en el telescopio la pulsar."

Pumpkin! Pumpkin! (POOM-keen POOM-keen), exp., suggesting that a party is over. Sp. *Calabaza Calabaza, todo el mundo pa' su casa.* {CA}

punchinbol (pon-cheen-BOLL), n., m., punching ball, solid sack boxers use for their training exercises. E»S

pundit (POON-deet), n., m/f., person accused of an offense. E»S

punk (PUNK), n., m., person who adheres to the style of the punk movement that started in Britain. E»S Also PUNKI.

pupilo (poo-PEE-lo), n., m/f., student. E»S

puritánico (poo-ree-TA-nee-ko), adj., puritanical. "Emily Dickinson era de un placete puritánico." E»S

puritano (poo-ree-TA-no). n., m/f., puritan, professing puritanical beliefs. E»S

pursa (POOR-sa), n., f., purse. "Le robaron la pursa a mi esposa." E»S

pushear (poo-SHEAR), v., to push. E»S Also PUSHIAR.

pushiar (poo-SHEAR), v., to push. E»S Also PUSHEAR.

put (POOT), n., m., 1. option to buy. "Tengo un put en ese apartamento." 2. small hit. Golf term. E»S

puter (poo-TER), n., m., person who makes a small hit. Golf term. E»S

puzle (PO-zleh), n., m., combination game, puzzle. Sp. *rompecabezas.* E»S Also POUZOL and POZEL.

quarc (KUARK), n., m., hypothetical nature element, part neutron, part proton.

quebradita (ke‑bra‑DEE‑tah), n., f., break, intermission. "Every day at school we have la quebradita at 11 AM." {Ch, SW}

quechear (ke‑CHEAR), v., to catch. "Ni quechea, ni pitcha, ni deja batear." E»S Also CACHAR and KECHAR.

quecher (KE‑cher), n., m., catcher.

quei (KEI), n., m., cake. "They like the quei de limón." {Ch, CA} Also QUEKI.

Queimar (KAI‑mahr), n., m., Kmart. "Voy d'compras at Queimar." {CA}

queis (KEIS), n., case. 1. legal case. "El abogado works en el queis de Juanito." 2. exp., in case. {Ch} Also CEIS and KEIS.

queki (keh‑KEE), n., m., cake. "They like the queki de limón." {CA, Ch} Also QUEI.

quesadilla (KE-sa-DEE-yah), n., f., tortilla with cheese. S»E

quidnapear (keed-na-PEAR), v., to kidnap. E»S

quidnaper (keed-NA-per), n., m/f., kidnapper. E»S

quiklista (KUEEK-lees-tah), n., f., commercial quick-list. {CS}

quina (KEE-nahh), n., f., 1. queen. "Ella es la quina de la casa." 2. fig-ure in cards. "I won the game con una quina." 3. police. "The quina is after the criminal." {Ch}

quiteado (KEE-teah-do), n., m/f., individual fired from the job.

quitear (KEE-tear), v., to quit. "Antonio quiteó su trabajo." E»S Also QUITIAR.

quitiar (KEE-tear), v., to quit. "Antonio quitió su trabajo." E»S Also QUITEAR.

R

rac (RAK), n., m., rack. "I hang my clothes in el rac." E»S

racin (RAY-seen), n., m., racing. E»S

racinguista (rah-seen-GEES-tha), n., m/f., athlete that practices racing.

racionalizar (ra-zio-na-lee-CZAR), v., to reflect. Sp. *reflexionar.* E»S

raftin (RAF-teen), n., m., rafting. E»S

rage (RAH-ge), n., m., popular music style from Jamaica.

ragtaim (rag-TAY-meh), n., m., a musical genre characterized by syncopated rhythm. E»S

rai (RAID), n., f., ride. "I waited por un rai por two hours." E»S Also RAID.

raid (RAID), n., f., ride. "I waited por un raid por two hours." E»S Also RAI.

raigrás (RAY-gras), n., m., field lawn. Eng. *rag grass.* "He sembrado raigrás in the backyard." {CA, Ch}

raite (RAI-teh), n., m., 1. automobile. "Me das un raite into town?" 2. Police raid. "La policía de Chicago hizo un raite en La Villita." E»S

raitin (REYH-teen), n., m., percentage of people that watch a TV show. E»S

ralí (rah-LEE), n., m., 1. automobile competition. Eng. *rally.* 2. ascendant action. Financial term. E»S

ramada (ra-MA-da), n., f., porch. The term dates from 1869. S»E

ranchero (ran-TCHE-roh), n., m/f., rancher. The term dates from 1826. S»E

rancho (RAN-tcho), n., m., 1. ranch. 2. type of salad dressing.

ranfla (RAN-flah), n., f., automobile. "My uncle has this hot ranfla with flames painted on it." {ELA} Also CARRUCHO.

rango (RAN-go), n., m., range. Sp. *rango* means rank.

rankin (RAN-keen), n., m., ranking list. E»S

rap (RAH-ph), n., m., music marked by a monotone rhythm.

rapeando (rah-PEAN-doh), adv., rapping. {Ch}

rapear (rah-pay-AHR), v. to rap. "Ese güe va a rapear in a club downtown." S»E

raquebol (ra-ke-BOLL), n., m., racketball. E»S

raqueta (ra-KE-tha), n., f., racket. From the Greek. E»S

raqueteada (ra-ke-TEAH-da), n., f., accusation for racketeering charges. E»S

raquetear (ra-ke-TEHAR), v., to racketeer. E»S

rascacielo (ras-ka-SEE-EH-loh), n., m., skyscraper.

rascuache (ras-KUA-tche), exp., 1. cheesy, of low quality. {Ch, Mex} 2. A person of inferior education. {Cen}

rasonar (ra‑zo‑NAR), v., to reason. Sp. *razonar.* E»S

raun (RAWN), n., m., 1. assault. 2. round. Boxing term. S»E

rayar (rah‑YAHR), v., to write. "Ella raya bien bonito for a doctor." {SW}

realístico (rea‑LEES‑ti‑ko), adj., realistic. Sp. *realista.* E»S

realitichou (re‑yal‑ee‑tee‑CHOU), n., m., TV show exploiting human miseries. E»S

realización (REA‑lee‑ZA‑zion), n., f., insight, realization. "I came to la realización that you're not even happy." E»S

realizar (rea‑lee‑ZAR), v., to achieve. "Juana realizó sus sueños through her studies." E»S

rebozo (REE‑bo‑zoh), n., m., 1. shawl. 2. sash.

rebú (ray‑BOO), n., m., dispute, quarrel. "They are still en su tonto rebú." {D}

rebutear (re‑boo‑TEHAR), v., to reboot.

recepción (re‑zep‑ZION), n., f., ceremony. E»S

recepcionar (re‑zep‑sio‑NAR), v., to organize a ceremony. E»S

recolección (re‑ko‑le‑XION), n., f., memory, reminiscence. "Después de ver la fotografía ella tuvo una recoleccón de su infancia." S»E

recolectar (re‑ko‑lec‑TAR), v., to remember. Sp. *recolectar* means to collect again. S»E

recón (re‑KON), v., to reckon. "Yo recón que ella doesn't know what she says." E»S

récor (REH‑cor), n., m., 1. measurement. "Angel tiene el récor of most trofeos in New Jersey." {CA} 2. v., to record on tape. 3. n., m., maximum result. E»S

recordear (reh‑kor‑DEAR), v., to record. E»S

récorman (REH-kor-man), n., m., person with a record. E»S

redear (re-DEAR), v., to be ready. Also used in the reflexive form. "Ella se va a redearse para la party." E»S

referencia (re-fe-REN-zia), n., f., 1. reference. 2. referral. "Necesito una referencia para conseguir el empleo." E»S

referí (re-fe-REE), n., m., person in soccer who arbitrates a match. Sp. *árbitro*. E»S

referir (re-fe-REER), v., to refer medically from one doctor to another. E»S

reflejar (re-fle-KHAR), v., to reason. Sp. *reflejar* means to reflect. E»S

reflex (REH-flex), n., m., photo-mechanism that allows an image to fix on the film.

refrainear (re-fray-NEAR), v., to refrain. E»S Also REFRAINIAR.

refrainiar (re-fray-NEAR), v., to refrain. E»S Also REFRAINEAR.

refri (REH-free), n., m., refrigerator. {SW}

refutar (re-foo-TAR), v., to refute, to oppose. "La madre de Luis refutó su argumento. No le permitío to go a la parti." E»S

registrar (re-gis-TRAR), v., 1. to enroll. 2. to certify. "Melissa se registró en la escuela." E»S

registro (re-GIS-tro), n., m., registry. "Hoy vamos al registro de automóviles." E»S

reguarda (reeh-GUAR-dah), n., f., chest protector made of leather used by a horse. "Dolores uses a reguarda por su accidente." S»E

reguardear (re-war-DEAR), v., to reward. E»S Also REGUARDIAR, REWARDEAR and REWARDIAR.

reguardia (re-WAR-dia), n., f., rewind. "Haz la reguardia para que escuchemos la primer canción del CD." E»S Also REWARDIA.

reguardiar (re‑war‑DEE‑ar), v., to reward. E»S Also REGUARDEAR, REWARDEAR and REWARDIAR.

regüín (re‑WEEN), n., m., mechanism by which a tape is rolled backwards. Eng. *rewind.* E»S Also REWIN.

regulación (re‑goo‑la‑ZION), n., f., regulation, law, code of conduct. E»S

reid (REYD), n., m., rapid military incursion. "El ejército norteamericano hizo un reid en la isla caribeña." E»S

reil (REYL), n., m., train rails. E»S

reinear (rei‑NEAR), v., to rain. E»S Also REINIAR.

reiniar (rei‑NEAR), v., to rain. E»S Also REINEAR.

relativo (re‑la‑TEE‑vo), n., m., family member. Sp. *relativo* means *relative.* "Pedro es un relativo de María. Sus padres son hermanos." E»S

relaxeao (reh‑lah‑kseh‑AH‑oh), adj., m/f., calm, subdued. "Es un tipo bien relaxeao." E»S

relaxar (re‑la‑XAR), v., 1. to rest. 2. relax. "Ay, qué día. Necesito relaxearme." E»S Also RELAXIAR.

relaxiar (re‑la‑XIAR), v., 1. to rest. 2. to relax. "Ay, qué día. Necesito relaxiarme." E»S Also RELAXEAR.

reliable (re‑LIA‑ble), adj., realiable. Sp. *reliable* means *confiable.* "Esa señora es reliable. Le puedes tener confianza." E»S

relise (re‑LEE‑seh), n., m., computer application that upgrades a previous one. {CS} E»S

relodear (re‑lo‑DEAR), v. to reload. {CS} E»S

rem (REM), n., m., term in physics that measures the radiation level. E»S

remarca (re‑MAR‑ka), n., f., 1. remark. 2. comment. 3. suggestion. E»S Also REMARQUE.

remarca (re-MAR-ka), n., f., remark. "El presidente hizo la remarca en la Casa Blanca." E»S

remarcable (re-mar-KA-ble), adv., remarkable. In Sp. *importante.* E»S

remarcar (re-mar-KAR), v., to remark. Sp. *remarcar* means to stress. E»S

remarque (re-MAR-ka), n., f., remark, comment, suggestion. Also RE-MARCA.

remedear (re-me-DEAR), v., to remedy. "Javier remedea su problema escolar con más horas en la clase." E»S Also REMEDIAR.

remediar (re-me-DEAR), v., to remedy. "Javier remedia su problema escolar con más horas en la clase." E»S Also REMEDEAR.

remeic (re-MEIK), n., m., fresh version of a preexisting product. E»S

remembrancia (re-mem-BRAN-zia), n., f., remembrance. "Ayer tuve una remembrancia de mi infancia." E»S

remembrar (re-mem-BRAR), v., to remember. E»S

remembroso (re-mem-BRO-zo), n., m/f., person with good mnemonic skills. E»S

remover (re-mo-VER), v., to remove. "Mi primo remueve su ropa del closet." E»S

rendear (ren-DEHAR), v., to render. "El músico rendea una canción de John Lennon en una nueva versión." S»E Also RENDIAR.

rendiar (ren-DEHAR), v., to render. "El músico rendia una canción de John Lennon en una nueva versión." E»S Also RENDEAR.

renegado (re-ne-GA-do), n., m/f., unrepentant outlaw. The term dates from 1599. "Gregorio Cortés era un renegado." Eng. *renegade.* E»S

replei (REH-play), n., m., program repeat on TV. "Vimos un episodio de *I Love Lucy* en un replei." E»S

reportar (re-por-TAR), v., to chronicle, to report. "Voy a reportar the situation." In Sp. *reportar* means to yield, to answer to, to call in sick. E»S

repórter (re-POHR-ter), n., m., journalist. "Me entrevistó un repórter del periódico." E»S

repós (re-POS), n., m., repurchase agreement. "Tengo el repós of the TV I bought at the store." E»S

reprografía (re-proh-gra-PHEE-a), n., f., reproduction of documents through mechanical means. E»S

resetear (re-seh-TEHAR), v., to reset. E»A

resignación (re-sig-na-ZION), n., f., resignation. "Hilda dio su resignación al jefe." In Sp. *renuncia* means resignation. E»S

resignar (re-see-GNAR), v., to hand in one's resignation. E»S

restar (res-TAR), v., to rest. "Necesito restar the game." In Sp. *restar* means to subtract. E»S

restorear (res-to-REAR), v., to restore. E»S

resurrectar (re-su-rrec-TAR), v., to resurrect, to bring back to life. E»S Also RESURRECTEAR and RESURRECTIAR.

resurrectear (re-su-rrec-TEAR), v., to resurrect, to bring back to life. E»S Also RESURRECTAR and RESURRECTIAR.

resurrectiar (re-su-rrec-TEAR), v., to resurrect, to bring back to life. E»S Also RESURRECTAR and RESURRECTEAR.

retalear (re-ta-LEAR), v., to retaliate. E»S Also RETALIAR.

retaliacion (re-talya-ZION), n., f., political or military retaliation. E»S

retaliar (re-ta-LEAR), v., te retaliate. E»S Also RETALEAR.

reteil (re-TAIL), n., m., retail business. E»S

retribución (re-tree-boo-ZION), n., f., retribution. E»S

retribuir (re-tree-boo-EER), v., to retribute. E»S

revaival (re-VAY-val), n., m., resurgence, recuperation. E»S Also RE-VIVAL.

reversar (re-ver-CZAR), v., to switch to the other side. E»S

revisionario (re-bee-zio-NA-ree-o), n., m., revisionist person. E»S

revival (re-VEE-val), n., m., resurgence, recuperation. E»S Also RE-VAIVAL.

rewardear (re-war-DEAR), v., to reward. E»S Also REGUARDEAR, REGUARDIAR and REWARDIAR.

rewardia (re-WAR-dia), n., f., reward. E»S Also REGUARDIA.

rewardiar (re-war-DEE-ar), v., to reward. E»S Also REGUARDEAR, REWARDIAR and REWARDEAR.

rewin (re-WEEN), n., m., mechanism by which a tape is rolled back-wards. Eng. *rewind.* E»S Also REGUIN.

rifer (REEH-fer), n., m., marijuana. {SW} Also GRIFA.

rift (REE-fth), n., m., 1. repeated musical phrase. 2. simple melodic style.

rim-an-blus (REEM-an-BLOOS), n., m., forties musical style. E»S

rímel (REE-mel), n., m., cosmetic used to color the eyelashes.

rín (REEN) n., m., 1. square-shaped arena. Boxing term. 2. rim of a wheel. E»S

rincón (reen-KON), n., m., nook, bend, secluded place. {Ch}

rines (REE-ayss), n., m., rims. {ELA}

riverski (REE-ver-SKEE), n., m., athletic activity that involves descend-ing through rough waters with skis and a paddle. E»S

robo (ROH-boh), n., m., automaton. Eng. *robot.* E»S

roc (ROC), n., m., also adj., rock 'n' roll. "Me encanta la musica roc, es-pecialtmente los Beatles." E»S

roca (ROH-kah) n., m., crack. From Eng. *rock*. "No hagas rocas if you don't want to be a drug addict." {Ch, Mex, SW}

rockabili (roh-KA-bee-lee), n., m., a type of rock 'n' roll from the southern US with influence of traditional songs. E»S

rocker (ROOH-ker), n., m., rock fan. E»S

rockero (roh-KEH-roh), n., m., rock musician. "Tengo un amigo rockero who is famous." E»S Also ROQUERO.

rodeo (ROOH-dee-o), n., m., 1. cowboy contest; 2. in colloquial Mexican Spanish, cattle roundup. In Sp. *rodear* means to surround.

rogby (ROOG-bee), n., m., rugby. Athletic activity. "Ellas juegan el rugby." E»S

rol (ROL), n., m., acting part. "Nicolás tiene el rol de Hamlet en la obra teatral." E»S

rola (ROOH-lah), n., f., fine song. {Ch}

rolar (roh-LAR), v., to roll, to cruise. "Estábamos rolando por Venice Boulevard when we saw Sal." {Ch, SW}

rooka (ROOH-ka), n., f., attractive woman. "Check it out! This' my rooka." {Ch} S»E Also RUCA.

roota (ROO-ta), n., f., root. E»S Also RUTA.

roguero (roh-KEH-roh), n., m., rock musician. "Tengo un amigo roguero who is famous." E»S Also ROCKERO.

rosbif (ros-BIF), n., m., roast beef. E»S

rotor (ROOH-tehr), n., m., rotting piece. Eng. *rotor*. E»S

roudcho (ROW-tcho), n., m., road show. E»S

rover (ROOH-ver), n., m., space exploration vehicle. {Ch}

royalti (roh-YAL-tee), n., m., person's rights in monetary measure. Eng. *royalty*. E»S

ruca (ROOH-ka), n., f., attractive woman. "Check it out, this' my ruca." {Ch} Also ROOKA.

rufa (ROOH-fah), n., f., roof. "Fix the rufa, está liqueando!" {CA, Ch, NR} Also RUFO.

rufo (ROO-fo), n., m., roof. "Fix the rufo, está liqueando!" {CA, Ch, NR} Also RUFA.

ruki (ROOH-kee), n., m., novice.

rumbo (ROOM-boh), n., m., 1. open space. Eng. *room.* {SW} 2. adj., plenty, sufficient. "That's enough already, my friend. I've eaten a rumbo." 3. Horses and carts might also be described as *rumbo,* good. "The horse is rumbo. It won second place in the Kentucky Derby. 4. n., m., closeness in geographical terms. "His house está en el rumbo." 5. adj., elegantly, fashionable by comparison. "Jaime is first-rate. Es del rumbo de Bel Air." {SW}

ruta (ROO-ta), n., f., root. E»S Also ROOTA.

S

S.O.S. (ESSE‑oh‑esse), n., m., salvo.

S.U.V. (SOOB), n., m., sports utility vehicle. E»S Also SUB.

sabe (SAH‑beh), exp., Do you know?

sabotear (sah‑boh‑TEAR), v., to sabotage. From the French.

sadlear (sad‑LEAR), v., to saddle a horse. "The cowboy sadlea el ca‑
 ballo." E»S Also SADLIAR.

sadliar (sad‑LEAR), v., to saddle a horse. E»S Also SADLEAR.

safo (SA‑pho), exp., May the Lord protect me! "Con safo." {SW} Also
 ZAFO.

safocón (zah‑pho‑CON), n., m., garbage container. {D, NY, PR} Also
 ZAFACON.

Sagüesera (sah‑weh‑SEH‑rah), n., f., southwestern section of Miami.
 {CA} E»S

saide (SAI-deh), n., m., siding, referring to the computer installer. {CS} E»S

sailear (say-LEAR), v., to sail. E»S Also SAILIAR.

sailiar (say-LEAR), v., to sail. E»S Also SAILEAR.

sainear (say-NEAR), v., to signal. "Martín sainea para que Lupe lo vea from afar." E»S

sala (sah-lah), n., f., dance hall.

salado (sah-LAH-doh), exp., wind-broken. Also SALAO.

salao (sah-LAH-doh), exp., wind-broken. Also SALADO.

salario (sa-LAH-rio), n., m., salary. E»S

salsa (SAAL-sah), n., f., 1. vegetable sauce. 2. Caribbean musical rhythm.

salvaguardar (sal-va-guar-DAHR), v., to safeguard. "Ella salvaguarda su dinero en el banco." E»S

salvar (sal-VAR), v., to save. {CS} E»S

sampler (sam-PLEHR), n., m., digital electronic equipment. {CS} E»S

sandgüichera (san-guee-TCHE-rah), n., f., sandwich maker. E»S

sandgüichería (san-guee-tche-REE-ah), n., f., sandwich shop. E»S

Sangívin (sahn-GEE-veen), n., m., Thanksgiving. E»S

sángüech (SAHN-wech), n., m., sandwich. E»S

sanguíneo (san-GEE-neoh), n., m/f., sanguine, self-confident. E»S

sanitación (sa-nee-TAH-tion), n., f., sanitation. E»S

sarái (sah-RYE), exp., It's alright! It's OK! "Sarái, don't worry! {CA, Ch}

sarcófago (sar-KOH-pha-go) n., m., meat-eater. {CA}

saundman (SAUN-man), n., m., sound technician. E»S

saundtrac (SAUN-trak), n., m., sound track. E»S

savi (SAH·vee), n., m/f., knowledge, wise. E»S

scolar (SKOH·lar), n., m/f., academic. E»S

scrach (SKA·tch), n., m., DJ technique to produce music. E»S

scrip (SCREEP), n., m., film text. E»S

scuash (SKUASH), n., m., squash. Athletic activity. E»S

secaucus (see·KAUH·cus), n., m., psychiatric hospital. {CA} E»S

secuoya (se·QUOH·ya), n., type of tree. Eng. *sequoia.*

sedán (she·DAN), n., m., standard automobile.

seekear (see·KEAR), v., to seek. E»S Also SEEKIAR, SIKEAR, SI·
KIAR, SIQUEAR and SIQUIAR.

seekiar (see·KEE·ar), v., to seek. E»S Also SEEKEAR, SIKEAR, SI·
KIAR, SIQUEAR and SIQUIAR.

segundo (se·GOON·doh), n., m., second·in·command in cattle busi·
ness. {SW} E»S

seguridades (se·que·ree·DA·des), n., f., securities. E»S

selectear (seh·lec·TEAR), v., to select. E»S Also SELESTAR and SE·
LECTIAR.

selectiar (seh·lec·TEE·ar), v., to select. E»S Also SELESTAR and SE·
LECTEAR.

selestar (seh·les·TAR), v., to select. E»S Also SELECTEAR and SE·
LECTIAR.

selestarter (se·leh·STAR·ter), n., m/f., person who selects. E»S

selfcontrol (self·kon·TROL), n., m., self·control. E»S

selfdefens (self·deh·FENZ), n., f., self·defense. E»S

selfevidente (self·e·vee·DEN·te), adj., self·evident. E»S

selfgovemen (self·go·ve·MEN), n., m., self·government. E»S

selfserviz (self·SER·viz), n., m., self·service. E»S

sénder (SEN-der), n., m/f., sender. {CS} E»S

Señor (she-NYOR), n., m., title of respect. The term dates from 1622. S»E

Señora (she-NYO-reah), n., f., title of respect. The term dates from 1579. S»E

sensitivo (sen-see-TEE-voh), n., m/f., sensitive. E»S

sénsor (SEN-zor), n., m., data transmitter. {CS} E»S

serape (she-RAH-peh), n., m., shawl worn as an outer garment. The term dates from 1834. S»E Also PONCHO

serendipiti (se-ren-DEE-pee-tee), n., f., accidental encounter. E»S

set (SET), n., m., portion of a tennis match. E»S

seter (SEH-ter), n., m/f., 1. fashion maker. 2. baby-sitter. E»S

setonly (SEAT-on-lee), exp., seat-only accommodation. E»S

sexapil (SEX-ah-peel), n., m., sexual attractiveness. E»S

sexi (seh-XEE), n., m/f., with sexual appeal. Also SEXY.

sexsho (SEX-shoh), n., f., sexual store. E»S

sexsimbo (sex-SEEM-bo), n., m., physically attractive star. E»S

sexy (seh-XEE), n., m/f., with sexual appeal. E»S Also SEXI.

shamac (shah-MAHKS), n., m/f., little kids. "The place was full of twelve-year-old shamacs." In Mex. Sp. *chamacos.* {Ch}

shantung (shan-TOON), n., m., Chinese silk.

sher (CHER) n., f., share, percentage. E»S

sherife (she-REE-pheh), n., m., sheriff. {NM, SW} E»S

shiftear (SHEEF-tehar), v., to shift. E»S

shipia (SHEE-pea-ah), n., f., shipyard. E»S

shoc (SHOK), exp., shock. E»S

shopin (SHOO‑peen), exp., buying. E»S

shopinba (sho‑peen‑BA), n., f., shopping bag. E»S

shopincenter (shoo‑peen‑CEN‑ter), n., m., commercial outlet. E»S

shora (SHO‑ra), n., f., shore. E»S

short (SHORT), n., m., short pants. Also used in plural. E»S

shotdaun (shot‑DAUN), n., m., automatic power shut‑down. {CS} E»S

shou (SHOW), n., m., show. E»S

shoubísnes (show‑BEES‑nes), n., m., show business. E»S

shouman (SHOW‑man), n., m/f., 1. presenter. 2. exhibitionist. "Javier es un verdadero shouman." E»S

shourum (SHOW‑room), n., m., theater. "Los actores están en el shou‑rum." E»S

sich (SEECH), n., m., side seat attached to motorcycle.

sichear (see‑CHEAR), v., to switch. E»S Also SICHIAR.

sichiar (see‑CHEE‑ar), v., to switch. E»S Also SICHEAR.

siesta (see‑ES‑tah), n. f., afternoon rest. The term dates from 1655. S»E

signatura (sig‑na‑TOO‑rah), n., f., signature. "The contract needs tu signatura." E»S

significante (sig‑nee‑fee‑KAN‑teh), n., m/f., significant other. "Achy vive con su significante en Chicago already for years." E»S

sikear (see‑KEAR), v., to seek. E»S Also SEEKEAR, SEEKIAR, SI‑KIAR, SIQUEAR and SIQUIAR.

sikiar (see‑KEE‑ar), v., to seek. E»S Also SEEKEAR, SEEKIAR, SI‑KEAR, SIQUEAR and SIQUIAR.

silbín (SEEL‑been), n., m., sealed beam. E»S

silicona (see‑lee‑KOH‑mah), n., m., silicone. E»S

Siliconvali (see‑lee‑kon‑VA‑lee), n., Silicon Valley. E»S

sílin (SEE-lean), n., m., ceiling. E»S

similaridad (see-mee-la-ree-DAD), n., f., similarity. E»S Also SIMI-LIDAD.

similidad (see-mee-lee-DAD), n., f., similarity. E»S Also SIMILA-RIDAD.

simón (see-MOHN), exp., yeah. "Did you finish? Simón, vámonos." {Ch, Mex} Also SINTURON.

simple (SEEM-pleh), n., m/f., simple-minded. E»S

singel (SEEN-gel), n., m/f., 1. unmarried. "Juán is forty-five years old but todavía está singel." 2. one-song recording number. "La canción de los Beatles está en un singel." E»S

sinturón (seen-too-RON), exp., yeah. "Did you finish? Sinturón, vámonos." {Ch, Mex} Also SIMON.

siquear (see-KEHAR), v., to seek. E»S Also SEEKEAR, SEEKIAR, SIKEAR, SIKIAR and SIQUIAR.

siquiar (see-KEE-ar), v., to seek. E»S Also SEEKEAR, SEEKIAR, SI-KEAR, SIKIAR and SIQUEAR.

sir (SEER), exp., 1. mister. 2. boss. "Yo trabajo para un sir muy exigente." E»S

skai (SKY), n. m., imitation-leather material used for upholstery. E»S

skailine (SKY-line), n., f., urban silhouette. E»S

skaitbord (SKATE-bord), n., m., skateboard. E»S

skaiter (SKATE-ehr), n. m/f., skateboard athlete. E»S

skech (SKETCH), n., m., sketch. E»S

ski (SKEE), v., to ski. "A los hermanos Martínez les gusta el ski." E»S

skinhed (SKEEN-hed), n., m/f., skinhead. E»S Also SQUINHED

slainear (slay-NEHAR), v., to slay. E»S Also SLAINIAR.

slainiar (slay-NEHAR), v., to slay. E»S Also SLAINEAR.

slango (SLAN-go), n., m., slang, jargon. "Nosotros hablamos un slango called Spanglish." E»S

slíper (SLEE-per), n., m., 1. sleeper [person]. 2. slow beginner [art]. "Esa película fue ignorada al principio pero terminó becoming un slíper." E»S

slogan (SLO-gan), n, m., commercial motto. E»S Also ESLOGAN.

smarticón (SMAR-tee-kon), n., m., smart icon. {CS} E»S

smash (SMASH), n., m., hit. "El último CD de Enrique Iglesias es un smash." E»S

smokin (SMO-keen), v., 1. to smoke. 2. n., m., tuxedo. E»S

smokinrum (smo-keen-ROOM), n., m., smoking room. E»S

snac (SNAK), n., m., snack. E»S

snacbar (SNAK-bar), n., m., snack bar. E»S

snob (SNOB), n., m/f., pretentious individual. E»S

snoubor (SNOW-bor), n., m., snow athletic item. E»S

snoubordear (SNOW-bor-dear), v., to snowboard. E»S

snouborder (SNOW-bor-der), n., m/f., snowboarder. E»S

snufmuvi (SNUFF-moo-vee), n., f., porno film.

sobretrot (so-bre-TROT), n., m., slow trot.

sócer (SO-ker), n., m., soccer.

social (so-TIAL), n., m., Social Security. E»S

soda (SOH-dah), n., f., soda, soft drink. Sp. *refresco.* E»S Also FRESCO.

sofisticado (so-fees-ti-KAH-do), n., m/f., artificial. In Sp. *sofisticado* means sophisticated. E»S

soflandin (sof-LAN-deen), n., m., soft airplane arrival. E»S

softwer (SOF-were), n., m., software. E»S

sokete (so-KE-te), n., m., sock. {SW} In Sp. *soquete* means socket. E»S Also SOQUETE.

sol (SOL), n., f., soul. "Esa gente no tiene sol." In Sp. *sol* means sun. E»S

soloista (so-lo-IS-tah), n., m/f., soloist. Sp. *solista.* E»S

sombrero (som-BRE-roh), n., m., Mexican hat. The term dates from 1770. S»E

són (SON), n., m., song. "Este són siempre makes me cry." E»S

sondáu (son-DAWN), n., m., sundown. E»S

sonóp (sun-UP), n., m., sunrise. E»S

soportar (so-por-TAR), v., to support financially. In Sp. *soportar* means to bear. E»S

soquete (so-KE-te), n., m., sock. {SW} In Sp. *soquete* means socket. Also SOKETE.

sororidad (soh-roh-ree-DAD), n., f., sorority. "Esas girls pertenecen a la sororidad de la university." E»S

sortear (SOR-tear), v., to sort. "Mauricio sortea el e-mail." {CS} E»S Also SORTIAR.

sortiar (SOR-tear), v., to sort. "Mauricio sortía el e-mail." {CS} E»S Also SORTEAR.

spaider (SPAY-der), n., m., sports automobile. "Dolores drives a gorgeous black spaider." E»S

spamear (SPA-mear), v., to spam. {CS} E»S

Spanglish (SPAN-gleesh), n., m., mestizo language, part English, part Spanish, used predominantly in the United States since WWII. Also CASTEYANQUI, GRINGONOL and INGLANOL.

Spanglization (span-glee-zah-TION), n., f., the process of Spanglish-language acquisition. E»S

sparin (SPA‑ren), n., m., boxing coach. E»S

sped (SPEED), n., m., speedy effect of drug stimulant. E»S

Spic (SPEEK), n., m., pejorative name for Latino. E»S Also SPICK and SPIK.

Spick (SPEEK), n., m., pejorative name for Latino. E»S Also SPIC and SPIK.

spich (SPEECH), n., m., speech. "El presidente dio un spich ante the audience." E»S

spicher (SPEE‑tcher), n., m/f., speaker. "Ramón González es un spicher." E»S Also SPIKER and SPIQUER.

Spik (SPEEK), n., m., pejorative name for Latino. E»S Also SPIC and SPICK.

spiker (SPEE‑ker), n., m/f., speaker. "Ramón González es un spiker." E»S Also SPICHER and SPIQUER.

spínin (SPEE‑neen), n., m., aerobics exercise. E»S

spínof (SPEEN‑of), n., m., division. "Ese programa de TV es un spinof del que mostraron el año pasado." E»S

spíquer (SPEE‑ker), n., m/f., speaker. E»S Also SPICHER and SPIKER.

splin (SLPEEN), n., m., spleen. E»S

split (SPLEET), n., m., split, division. E»S

spoiler (SPOY‑lehr), n., m/f., party disturber. E»S

spónsor (SPON‑sor), n., m., endorser. E»S

sponsorizar (spon‑zo‑ree‑ZAR), v., to sponsor. "Coca‑Cola sponsoriza the holiday race." E»S

spor (SPOR), n., m., sport. "Ellos son aficionados al spor." E»S

sporguer (SPOR‑wear), n., m., sportwear. E»S Also SPORWER.

sporsman (SPORS‑man), n., m., sport person. E»S

sportin (SPOR‑teen), n., m., sporting equipment. E»S

sporwer (SPOR‑wear), n., m., sportwear. E»S Also SPORGUER.

spot (SPOT), n., m., 1. occasion. 2. opportunity. 3. TV commercial. E»S

spotlait (SPOT‑lait), n., m., public attention. "Juana recibió el spotlait after her first novela was published." E»S Also SPOTLIGHT.

spotlight (SPOT‑lait), n., m., public attention. "Juana recibió el spot‑light after her first novela was published." E»S Also SPOT‑LAIT.

spred (SPRED), n., m., solid advertisement. E»S

sprei (SPRAY), n., m., spray. E»S

sprin (SPREEN), n., m., spring season. E»S

sprint (SPREENT), n., m., rush. E»S

sprintear (SPREEN‑tear), v., to sprint. "La atleta sprintea al final of the race." E»S

sprinter (SPREEN‑ter), n., m., sprinter. "Ese atleta es el sprinter más rápido." E»S

spufear (spoo‑FEAR), v., to spook. "Isaías spufea a su familia con cuen‑tos de horror." E»S

spufeo (spoo‑FE‑oh), n., m., ugly spook. E»S

squinhed (SKEEN‑hed), n., m/f., skinhead. E»S Also SKINHED

staf (STAF), n., m., staff. E»S

staig (STEI‑gh), n., m., stage. "Kenneth Branagh está en el staig en Broadway." E»S

stampede (STAM‑peed), n., f., sudden cattle panic. The term dates from 1828. S»E

stand (STAND), n., m., position, stand. "Ese stand que tomó la Unión les costó la huelga." E»S

standar (STAN-dar), n., m., standard. E»S

standbai (STAND-bay), n., m., stand-by. E»S

standin (STAN-deen), n., m., position within chart. E»S

star (STAR,), n., m/f., celebrity. "Esa niña sueña con ser una star." E»S

starleta (star-LE-ta), n., f., second-rate female star. "Juana Méndez es una starleta de películas gringas." E»S

starter (STAR-ter)., n., m., ignition mechanism. E»S

stationario (sta-tio-NA-rio), n., m., stationery paper. "The secretary ordenó stationario for the office." E»S

stedicam (ste-dee-KAM), n., m., stable-camera images." Ellos usan la stedicam en el viaje a Nigeria." E»S

steitofdeart (STEIT-of-deh-ART), n., m., state of the art. E»S

sténcil (STEN-seel), n., m., translucent paper. E»S

step (STEP), n., m., dance act. "Aprendimos un nuevo step en la clase de danza." E»S

stéreo (STEH-re-oh), n., m., stereo. E»S

stic (STEEK), n., m., hockey stick. "Los jugadores tienen nuevos stics." E»S

stilo (STEE-loh), n., m., style. "Sorry, vato, I didn't know that was your stilo."

stim (STEEM), n., m., steam. E»S

stoc (STOK), n., m., stock. E»S

stocmarke (STOK-mar-keh), n., m., stock market. "Perdimos dinero en el stocmarke." E»S

stocopcion (STOK-op-tion), n., m., stock option. "Estamos atentos para conseguir nuevas stocopciones." E»S

stop (STOP), n., m., traffic signal. "Llegamos al stop." E»S

stoplos (STOP-los), n., m., sudden loss of recognition. "In the accident, Juan tuvo un stoplos." E»S

stoque (STO-keh), n., m., stroke. E»S

storibor (sto-ree-BOR), n., m., story board. E»S

straic (es-TRAIK), n., m., strike. Athletic term. "El bateador dio un straic." E»S Also ESTRAIK and STRAIK.

straik (es-TRAIK), n., m., strike. Athletic term. "El bateador dio un straik." E»S Also ESTRAIK and STRAIC.

stréchin (STRE-cheen), n., m., muscular stretch. E»S

strékin (STRE-keen), n., m., nude protest. "Los estudiantes hicieron un strekin en el campus para protestar la guerra." E»S Also STRE-QUIN.

stréquin (STRE-keen), n., m., nude protest. "Los estudiantes hicieron un strequin para protestar la guerra." E»S Also STREKIN.

stres (STRES), n., m., stress. E»S Also ESTRES.

stresante (streh-SAN-teh), exp., stressing. "Esta situación está stresante." E»S

streseado (stre-SEHA-do), n., m/f., stressed out. E»S

stresear (STRE-sear), v., to stress. E»S Also ESTRESAR.

strik (STREEK), n., m., streak. E»S

striper (STREE-per), n., m./f., stripper. E»S

striptis (streep-TEES), n., m., striptease. E»S

suap (soo-AP), n., m., exchange. "El cassette está defectuoso. Queremos hacer un suap." E»S

sub (SOOB), n., m., sports utility vehicle. E»S Also S.U.V.

subcontinental (sub-kon-tee-nen-TAL), n., m/f., individual with origins in India and Pakistan. "El profesor de matemáticas es subcontinental. Viene de Paquistán." E»S

subcontinente (soob·kon·tee·NEN·te), n., m., subcontinent. "Sahar Ahmed es del subcontinente. Ella nació en India." E»S

subcultural (soob·kul·too·RAL), n., m/f., undereducated. E»S

submitir (sub·mee·TEER), v., to submit. "El director submite su renuncia." E»S

submitirse (sub·mee·TEER·se), v., to surrender. "After committing the crime, Manuel Pedraza se submitió a la policia." E»S

suburbial (soo·bur·BIAL) n., m., suburbs resident. E»S

suceso (soo·XEH·soh), n., m., success. E»S

sucesor (soo·XEH·sohr), n., m/f., successor. E»S

sudenmente (soo·den·MEN·te), adv., suddenly. E»S

sueñoso (sweh·NYO·so), n., m/f., sleepy individual. E»S

suéter (SWEH·ter), n., m., sweater. E»S

sugestionar (soo·ges·tio·NAR), v., to persuade. E»S

suich (soo·ITCH), n., m., 1. electric device. 2. exchange. E»S Also SUICHE.

suiche (SWEE·cheh), n., m., 1. electric device. 2. exchange. E»S Also SUICH.

suing (soo·EENG), n., m., 1. curved movement. "El beisbolista hizo un suing con el bat." 2. thirties music style. E»S

sujeto (soo·KHE·to), n., m., theme, subject. "I don't understand el sujeto de esta novela." In Sp. *sujeto* means subjected to. E»S

sumarizar (soo·mah·ree·ZAR), v., to summarize. E»S

sumarizarse (soo·mah·ree·ZAR·se), v., to sum oneself up to a group or occasion. E»S

sumisivo (soo·mee·SEE·voh), n., m/f., submissive. E»S

súper (SOO·per), 1. n., m., supermarket. "Voy al súper, do you want anything?" 2. building superintendent. "El súper doesn't want to fix the door." 3. adv., superlative. "This new house está súper buena." E»S

superego (soo-per-EH-goh), n., m., self. "María tiene un superego vulnerable."

supermán (soo-per-MAN), n., m., superior person. "Lisandro es un superman. Puede levantar un automóvil con los dos brazos." E»S

supermojado (soo-per-mo-KHA-doh), n., m/f., border hero. E»S

supervisar (soo-per-vee-ZAR), v., supervise. E»S

supervisión (soo-per-vee-ZION) n., f., supervision. E»S

supervisor (soo-per-VEE-zor), n., m/f., supervisor. E»S

superwoman (soo-per-WHO-man), n., f., superior person. "Nancy es una superwoman. Trabaja 45 horas a la semana." E»S

supórter (soo-POR-ter), n., m/f., supporter. E»S

surf (SOORF), n., m., surf. "Está en el surf." E»S

surfear (SOOR-fear), v., to surf. "Ese güey no quiere ir a la escuela. He just wants to surf." E»S

surfer (SOOR-fer), n., m/f., surfer. E»S Also SURFISTA.

súrfin (SOOR-feen), n., m., nautical sport. E»S

surfista (soor-FEES-tah), n., m/f., surfer. E»S Also SURFER.

suspenso (sus-PEN-so), n., m., suspense.

sustitutivo (sus-tee-too-TEE-voh), n., m., replacement. "Rodrigo pidió un sustituvo at work porque he is sick." E»S

sute (SOO-teh), n., m. suit. E»S

T

ta (TAH), n., m., sales tax. "I thought I had enough money, pero olvi-
daba lo del ta. E»S Also TAKS.

tabloide (tah-BLOY-dee), n., m., tabloid newspaper. "Leímos la noticia
en el tabloide." From Latin.

taca (TAH-cah), n., f., tattoo. "Oye, how many tacas do you have?"
{Ch, SW}

tachar (tah-CHAR), v., to touch. "Táchalo, carnal. You'll see how soft
it is!" In Sp. *tachar* means to cross out. {Ch}

tag (TAAG), n., m., graffiti symbol. E»S {Ch, SW}

taguear (tah-GUEAHR), v., to tag. E»S

taibreik (TAI-braik), n., m., even score. Tennis term. E»S

taifa (TAY-fah), n., m., thief. {Chi}

taikover (TAYK-over), n., m., hostile corporate take-over. "Mi compañía
fue víctima de un taikover." E»S

taimin (TAY‑meen), n., m., calendar. Eng. *timing.* E»S

taimquiper (TAIM‑kee‑per), n., m., timekeeper. E»S

taipear (tay‑peh‑AHR), v., to type. E»S Also TAIPIAR.

taipfaiz (TAYP‑feiz), n., m., typeface. {CS}

taipiar (tay‑peeh‑AHR), v., to type. E»S Also TAIPEAR.

taipo (TAY‑poh), n., m., error. E»S

taks (TAX), n., m., tax. E»S Also TA.

talismán (ta‑LEES‑man), n., m., amulet. The term dates from 1638. "The cowboy keeps his talisman for good luck." S»E

tamale (tah‑MAH‑leh), n., m., corn husk. The term dates from 1856. S»E

tampax (TAM‑pax), n., m., hygienic tampon, brand name. E»S

tampón (tam‑POOHN), n., m., tampon. E»S

tañado (tah‑NYAH‑doh), n., m/f., tanned. E»S

tangibilizar (tan‑gee‑bee‑lee‑ZAR), v., to make tangible. E»S

tango (TAN‑goh), n., m., South American dance. The term dates from 1896. S»E

tanque (TAN‑keh), n., m., swiming pool. E»S

tanquear (tan‑KEAR), v., to swim.

tantalizar (tan‑tah‑lee‑ZAR), v., to tantalize. E»S

tarifa (rah‑REE‑fah), n., f., price. "La tarifa de la TV es $275." Sp. *tarifa* means tarif.

tarjeta (tar‑KHE‑tah), n., f., 1. objective, goal. Eng. *target.* "We shall achieve la tarjeta." {CH} 2. credit card. "Usa la tarjeta for the purchase." Also PLASTICO.

tasco (TAS‑koh), n., m., task. E»S

taunship (TAUN·ship), n., m., South African republic inhabited by people of color. E»S

tecato (te·KAH·teh), n., m., beer, brand name. {Ch} E»S

tecletear (the·kleh·TEAR), v., to type. E»S

tecleteo (teh·kleeh·TEH·oh), n., m., the typing sound. E»S

tecno (TEK·noh) n., m., techno music. "No puedo bailar al tecno, I pre· fer rock 'n' roll. {CS} E»S

teddy·boy (THE·dee·boi), n., m., sixties youth with loose pants. {Ch} E»S

teilor (TEY·lohr), n., m., tailor. E»S

teip (TAYP), n., m., tape. E»S

teipear (TAY·peh·AR), v., to tape record. E»S

teleconferencia (the·leh·con·phe·REN·ziah), n., f., phone conference.s E»S

telefón (the·leh·PHON) n., m., telephone. E»S

teletext (te·LEE·text), n., m., TV transmission of information. {CS} E»S

teletipo (the·leh·TEE·poh), n., m., teletypesetter. E»S

telex (TEE·lyx), n., m., International Telegraphic System. E»S

templait (ten·PLAYT), n., m., template. E»S

tenan (TEN·an), n. m/f., tenant. E»S

tenis (TEH·nees), n., m., 1. sneakers. 2. tennis. "Put on los tenis." E»S

tensén (ten·SEN), n., m; a ten·cent store, five·and·dime. "Si nada más quieres dulces, it would be best to go to the tensén on the cor· ner. E»S

terminador (ter·mee·NAY·tohr), n., m., violent fighter. "Javier es un terminador. Los vatos le tienen miedo." {CS} E»S

tes (TEHS), n., m., test. E»S

tester (TES-ter), n., m., taste maker. E»S

tex (TEKS), n., m., text. "Ella terminó el tex pa' la clase de francés." E»S

ticher (TEE-chair) n., m., teacher. E»S

tifiar (tee-fee-AHR), v., to steal. {C, CA}

tilburi (teel-buh-REE), n., m., uncovered carriage for two people pulled by a horse. Eng. *Tilbury.* E»S

tilde (TEEL-deh), n., f., diacritical sign over n to indicate palatalization. The term dates from 1864. S»E

timba (TEEM-bah), n., f., large piece of wood. {CA}

timin (TEE-meen), n., m., act of programming dates or events. Eng. *timing.*

tineijer (teen-AI-ger), n., m/f., teenager. E»S

tip (TEEP), n., m., tip. E»S Also TIPA.

tipa (TEE-pa), n., m., tip. E»S Also TIP.

tipear (tee-PEHAR), v., 1. to tip. 2. to type. E»S Also TAIPEAR.

tipificar (tee-pee-fee-KAR), v., to individualize. E»S

tique (TEE-keh), n., m., ticket. "Tengo un tique más in order to go to the concert."

tiquear (TEE-kear), v., 1. to issue tickets. 2. To tick, as in a clock. E»S Also TIQUIAR.

tiquete (tee-KEH-teh), n., m., ticket. "Aqui tengo tu tiquete for the concert. Here's your ticket for the concert." {CA, NY, PR} E»S Also TIQUE.

tiquetero (tee-KEH-the-ro), n., m., ticket seller. E»S

tiquiar (TEE-kee-ar), v., 1. to issue tickets. 2. to tick, as in a clock. E»S Also TIQUEAR.

tishu (TEE-shoe), n., m., tissue paper. E»S

tiviri (TEE-vee-ree), adj., m/f., f., 1. activity. 2. socially busy. Eng. *activity*. {C, CA} E»S Also ESTAR DE TIVIRI.

tobacco (toh-BAH-ko), n., m., narcotic dried leaves. The term dates from 1577. S»E

tobogán (toh-boh-GAN), n., m., toboggan. E»S

tocsho (TOK-shoh), n., m., talk show. E»S

tofi (TOH-fee), n., m., bland caramel made of coffee with milk or chocolate. Eng. *toffee*. E»S

toile (TOY-leh), n., m., toilet. Fr. *toilet*.

tokear (toh-KEAR), v., to talk. E»S

tolerar (to-le-RAR), v., to tolerate. Sp. *aguantar*. E»S

tomar acción (toh-MAR ah-xion), exp., take action. E»S

tomar efecto (toh-MAR eh-FEK-toh), exp., take effect. "The law toma efecto this month." E»S

tomar examen (toh-MAR eh-XA-men), exp., take a test. "Esther toma examen." E»S

tomar excepción (toh-MAR ex-zep-ZION), exp., take exception. "Ella tomó excepción to the general opinion in the classroom." E»S

tomar noticia (toh-MAR no-TEE-zia), exp., take notice. "Toma noticia de la enfermedad que sufre tu madre."

tomar riesgo (toh-MAR ree-EZ-goo), exp., take a risk. "Esteban tomó el riesgo and he lost his money." E»S

tomato (toh-MAH-toh) n., m., red garden vegetable. The term dates from 1604. S»E

tonape (toh-NAH-peh), n., m., turnip. {NM}

tóner (TOH-ner), n., m., toner. E»S

toner (TOH-ner), n., m., ink used in laser printer. "La impresora nece-sita toner." S»E

tonsila (ton-SEE-lah), n., f., tonsils. E»S

top (TOHP), n., m., item of feminine clothing. E»S

top model (TOHP MO-del), n., f., in-style person. E»S

topless (tohp-LESS), n. m., naked female breasts. E»S

topsic (TOHP-seek), adv., top secret. E»S

toptén (tohp-TEN), top ten, Top Ten list. E»S

torero (toh-REH-roh), n., m., bullfighter. The term dates from 1728. S»E

tori (TOH-ree), n., m., conservative British politician. E»S

tortilla (tohr-TEE-yah), n., f., cornmeal pancake. The term dates from 1699. S»E

tostado (TOUH-st), n., m., toasted bread. E»S

totalizar (toh-tah-LEE-zar), v., to complete. E»S

traca (TRAH-kah), n., m., railroad track. {Mex, SW} E»S

trail (TRAY-el), n., m., ability test. E»S

trailear (tray-LEHAR), 1. n., m., motorcycle sport. 2. v., to wander around on a truck. {SW} E»S Also TRAILIAR.

trailer (TRAY-leer), n., m., 1. tow truck; 2. movie preview. 3. suspense film. Eng. *thriller.* E»S

trailera (TRAY-le-ra), n., f., loose woman encountered on the road. {Ch, Mex, SW} E»S

trailiar (TRAY-lee-ar), 1. n., m., motorcycle sport. 2. n., m., to wander around on a truck. {SW} E»S Also TRALEAR.

trainin (TRAY-neen), n., m., practice period. E»S

tramados (tra-MA-dos), n., m., pants. {Ch}

tramp (TRAMP), n., m., shipping boat. E»S

transfer (TRANS-fer), n., m., soccer term that describes the transfer of players from one team to another. E»S

transistor (tran-sees-TOHR), n., m., electronic equipment that amplifies sound.

transpasar (tras-pah-ZAR), v., to trespass.

tranvía (tran-VEE-a), n., m., tramway. E»S

traque (TRAH-kee), n., m., track. E»S

trash (TRA-sh), n. f., 1. garbage. 2. accelerated musical rhythm. "She enjoys dancing la trash." E»S

tratar (trah-TAR), v., to treat, to handle. "Ella trata a su amiga without respect." E»S

travelers (TRA-vee-LERS), n., m., energy transmitted through cables. E»S

travelerscheks (TRAH-veh-lers-tcheks), n., m., tourist checks. E»S

travelín (TRA-vee-LEEN) n., m., in film, video, or TV, horizontal movement which the camera shoots from the top of a platform to film the movement of a round plane. E»S

treider (TRAI-der), n., m., professional dedicated to be an intermediary between individuals or companies. Eng. *trader*. E»S

treidin (TRAI-deen), n., m., transaction. E»S

treidmark (TRAID-mark), n., f., registered brand. E»S

trequin (TREE-keen), n., m., sport that involves walking through diffi-cult terrain. E»S

tribal (tree-BAL), n., m., primitive, belonging to a tribe. "Esta situation es tribal." E»S

trip (TREE-ph), n., m., hallucinogenic dose. "Los trips son acid dreams." {Ch, NR, S, SW} E»S

tripear (tree‑peh‑AHR), v., to experience abnormal states of mind. Eng. *trip.* {Ch, NY, S} E»S

tripeo (tree‑PEH‑oh), n., m., an unusual or mind‑bending experience. {Ch, NY, S} E»S Also TRIPEAR.

tripioso (tree‑pee‑OH‑soh), adj. Uncanny, mind‑shattering. In slang Eng. *trippy.* {Ch, NY, S} E»S

troca (TROH‑kah), n., f., truck. "Jorge manejó la troca but he doesn't have a license." {M, SW} Also TROKA.

troka (TROH‑kah), n., f., truck. "Jorge manejó la troka but he doesn't have a license." {M, SW} Also TROCA.

tronco (TRON‑koh), n., m., trunk. E»S

troquero (troh‑KEH‑roh), n., m., truck driver. {M, SW} E»S

trucha (TROO‑chah), adj., alert. "Don't try to fool her, she's bien trucha." {ELA}

truismo (troo‑EEZ‑moh), n., m., truism. From Eng. *true.* E»S

trust (TROS‑th), n., m., partnership. of companies. E»S

tuchdaun (toch‑DAWN), n., m., touchdown. Football term. E»S

tudú (too‑DOO), n., m., chore, to‑do. "Tengo un tudú tonight." E»S

tuid (TWEED), n., m., multicolor wool cloth. E»S Also TWEED.

turbolancha (toor‑boh‑LAN‑tcha), n., f., turbo boat. E»S

turf (TOR‑ff), n., m., race track. E»S

turismo (TOO‑rees‑MOH), n., m., 1. tourism; 2. private vehicle with 4 or 5 seats. E»S

tweed (TWEED), n., m., multicolor wool cloth. E»S Also TUID.

twinset (TWEEN‑set), n., m., women's clothing made up of shirt and jacket. {S} E»S

twist (TWEE‑sth), n., m., dance style. E»S

U

U.F.O. (OO‑fo), n., m., Unidentified Flying Object. Sp. *O.V.N.I.* "Ayer vimos un UFO in the sky" E»S

uantutri (uahn‑too‑TREE), n., m., immediately. E»S

ufología (oo‑fo‑lo‑GEE‑a), n., f., ufology, discipline devoted to studying UFOs. E»S

újole (OO‑kho‑lee), interj., denoting surprise, ridicule. "Újole, ya se lo llevaron. Wow, he was taken away." {M, SW} E»S

Unaited Esteits (u‑NAY‑ted es‑TAYTS), n., m., United States. E»S

unalienable (un‑a‑lehan‑A‑ble), adj., protected. "These derechos are un‑alienables." E»S

undear (on‑DEHAR), v., to undo. E»S

undeletear (on‑DEH‑le‑tear), v., to rescue a deleted text or document. {CS} E»S

understear (on-der-STEAR), v., to understand. E»S

unión (U-knee-ON), n., f., worker's union. E»S

unión de canerías (uh-NYON deh ka-ne-REE-has), n., f., cannery
workers. {Chi}

upgrade (UP-grayd), n., m., upgrade. "Necesito un upgrade." E»S

upside (UP-sayd), v., to ascend. E»S

urgir (URH-geer), v. to urge. E»S

uta (OO-tah), interj., Damn! "¡Uta! Damn, he's gone." Sp. *puta madre.*
{M, SW} S»E

utilidad (u-tee-lee-DAD), n., f., utility, tool. Pl. *utilidades.* "This ham-
mer is la utilidad que necesitas para construir el mueble." E»S

utiliti (yoo-TEE-lee-teah), n., m., utility room. {CA} E»S

V

vacada (va·KAA·da), n., f., cow drove. S»E {SW, M}

vacunar (ba·koo·NAR), v., to vacuum. "Mi hermana vacuna the carpet."
E»S Also VACUNEAR and VAQUIMEAR.

vacuncliner (va·koon·KLEE·nair), n., m., vacuum cleaner. {CA, NY,
SW} E»S

vacunear (va·koo·neh·AHR), v., to vacuum. E»S Also VACUNAR and
VAQUIUMEAR.

valorizar (va·loh·ree·ZAR), v., to value. E»S

valuable (vah·LUAH·bleh), n., m., valuable. "Los valuables are in the
safe deposit box." E»S

valuación (vah·LUAH·tion), n., f., act of establishing the value of an
item or event. "La valuación de la fiesta es falorable." E»S

vamus (va·h·MOO·see), exp., Let's go! Get out of here! "Vamus, vato.
Let's go, brother." {Ch, SW} E»S

vaquero (vah‑KEh‑roh), n., m., 1. cattle rancher; 2. jeans. "That old va‑ quero will never leave the range." Also BUCKAROO and COWBOY. S»E

vaquiumear (va‑keeu‑MEAR), v., to vacuum. Also VACUNAR and VACUNEAR. E»S

variancia (va‑reean‑ZEAH), n., f., variance. E»S

vate (VA‑tee), n., m., water. {Ch} E»S

vato (VA‑toh), n., m., 1. acquaintance of Chicano descent; 2. dude, homeboy; 3. Chicano gangbanger. "Orale vato! Hey, man, what's happening!" {Ch, ELA} Also BATO, BRO, BRODER, BROTHER, GÜE and GÜEY.

vegetales (veh‑ge‑TA‑les), n., m., vegetables. "Today we eat carne con vegetales." E»S

veigo (VAI‑goo), v., I see. From Sp. *veo*. {SW}

veintidós (VAIN‑tee‑dos), n., f., .22‑caliber rifle. {SW}

velados (Ve‑LAH‑dos), n., m., newlyweds. {SW}

velís (ve‑LEES), exp., Heads up! "Estate velís. Be on the lookout!" In Sp. *velís* means suitcase. {SW}

verdadera bosa (vair‑dah‑DEH‑rah BOH‑sah) n., f., real boss. "Mi mujer es la verdadera bosa. My wife is the real boss." {CA}

vernacular (ver‑NAH‑coo‑lar), n., m., vernacular. "Ellos speak en el ver‑ nacular." E»S

versátil (ver‑SAH‑teel), n., m/f., talented. "She es an actress versátil." In Sp. *versátil* means agile. E»S

Viaypi (vee‑AY‑p), n., m/f., individual of high social standing, V.I.P. E»S

víbora (VEE‑bo‑rah), n., m., penis. {Ch}

videasta (vee‑dee‑HAS‑tah), n., m/f., video professional.

videocaset (vee·deo·ka·SET), n., m., videocassette. E»S

videoclip (VEE·dee·o·KLEEP), n., m., brief video recording. E»S

videoclub (vee·dee·o·CLUB), n., m., video club. E»S

videoconference (vee·dee·o·KON·fe·rence), n., f., video conference. E»S

videometraje (vee·deo·me·TRA·khe), n., m., video movie. E»S

videoplais (vee·dee·o·PLAYZ), n., m., 1. audiovisual effect using com·puter·generated images. 2. video store. E»S

videopub (vee·dee·o·PUB), n., m., video pub. E»S

videotaip (vee·dee·o·TAIP) n., m., video tape. E»S

videotext (VEE·dee·o·TEHS), n., m., TV or phone informational. E»S Also VIDEOTEXTO.

vidio (VEE·dee·ho), n., video. E»S

vigilante (vee·gee·LAN·tee), member of vigilance committee. The term dates from 1856. S»E

vilificar (vee·lee·fee·KAR), v., to vilify. E»S

villita (vee·YEE·tah), n., m., neighborhood. {SW}

violación (vio·la·TION), n., f., infraction, breach, transit violation. In Sp. *violación* is often used for rape. E»S

virtual barrio (VEER·tu·al BA·ree·o), n., m., World Wide Web page devoted to Chicano issues. {CS, Ch, SW} Also BARRIO VIRTUAL.

virula (VEE·ru·lah), n., f., bicycle. {SW}

visualizar (vee·ZUA·lee·zar), v., to visualize. E»S

voismeil (VOYS·mel), n., m., voicemail. E»S

voleibol (vo·LEI·ball), n., m., volleyball. E»S

W

W.A.S.P. (WASP), n., m/f., White Anglo-Saxon Protestant.

wacha (GUA-tcha), exp., watch out. E»S Also GUACHA and WA-CHALE.

wáchale (WA-cha-leh), exp., 1. to watch. 2. to look after. "Wáchale, carnal, allí viene la Migra. Watch out!" {Ch, SW} Also GUACHA and WACHA.

wachear (WA-tchear) v., 1. to observe. 2. to watchout. Also GUA-CHEAR and GUACHIAR.

wachimán (wa-chee-MAN), n., m., guard. {Cen, Ecuador, Venezuela} E»S. Also GUACHIMAN.

wafle (WA-fle), n., m., waffle. E»S Also GUAFLE.

waflear (WA-flear), v., to eat waffles. E»S Also GUAFLEAR.

wagonlit (wa-gon-LEET), n., m., wagon lit. E»S

waipe (WAI-peh), n., m., windshield wiper. E»S Also GUAIPE.

waitear (way-TEHAR), v., to wait. E»S. Also WAITIAR, WEITEAR and WEITIAR.

waitiar (way-TEEHAR), v., to wait. E»S. Also WAITEAR, WEI-TEAR and WEITIAR.

waki (WA-kee) n., m., walkie-talkie. {Cen, Mex, S} E»S

wakman (WALK-man), n., f., walkman, music box.

waksear (wah-KSEAR), v., to wax. {NY} E»S

walkin (WA-keen), adv., walking at a fast speed. E»S

wandear (wan-DEHAR), v., to wander. {CS} E»S

waract (war-ACT), n., m., act of war. E»S

warmop (war-MOP), n., m., athletic warm-up. E»S

warran (WA-rran), n., f., stock bond with a period fixed rate. {CA, DA} E»S

wasá (wa-SA), exp., 1. what's up? 2. what's happenning? "Wasá, hombre, how are you?" 3. n., f., joke. "Maribel hizo una wasá que ne hizo reir." {Ch, SW} E»S Also GUSA.

washatería (wa-tcha-te-REE-ah), n., f., laundry store. E»S

wateado (wa-TEA-do), adj., m/f., watered. "Esa ropa está wateada."

watear (wa-TEHAR), v., to wet. E»S

watercloset (wa-ter-CLOH-set), n., m., toilet. E»S

waterpolista (wa-ter-po-LIS-ta), n., m., water polo athlete. {S} E»S

waterpolo (wa-ter POH-loh), n., m., water polo sport. E»S

waterpú (wa-ter-POO), n., m., water polo. {S} E»S

weakenear (we-a-ke-NEHAR), v., to weaken, to debilitate. E»S Also WEAKENIAR, WEAQUENEAR, WEAQUENIAR, WI-KENEAR, WIKENIAR, WIQUENEAR and WIQUE-NIAR.

weakeniar (we-a-ke-NEHAR), v., to weaken, to debilitate. E»S Also WEAKENEAR, WEAQUENEAR, WEAQUENIAR, WI-KENEAR, WIKENIAR, WIQUENEAR and WIQUE-NIAR.

weaquenear (we-a-ke-NEHAR), v., to weaken, to debilitate. E»S Also WEAKENEAR, WEAKENIAR, WEAQUENIAR, WIKE-NEAR, WIKENIAR, WIQUENEAR and WIQUENIAR.

weaqueniar (we-a-ke-NEHAR), v., to weaken, to debilitate. E»S Also WEAKENEAR, WEAKENIAR, WEAQUENEAR, WI-KENEAR, WIKENIAR, WIQUENEAR and WIQUE-NIAR.

websait (web-ZAIT), n., f., web site. {CS} E»S

weioflaif (way-of-LAIF), n., m., way of life. E»S

weitear (way-TEHAR), v., to wait. E»S Also WAITEAR, WAITIAR and WEITIAR.

weitiar (way-TEEHAR) v., to wait. E»S Also WAITEAR, WAITIAR and WEITEAR.

welfaresteit (wel-fer-es-TAIT), n., m., welfare state. E»S

welfér (well-FER), n., m., welfare system. E»S

wermut (wer-MOOTH), n., m., vermouth. E»S

western (WES-tern), n., m., cowboy movie. E»S

wieldear (weel-DEHAR), v., to wield. E»S Also WIELDIAR.

wieldiar (weel-DEHAR), v., to wield. E»S Also WIELDEAR.

wikén (wee-KEN), n., m., weekend. E»S

wikenear (wee-ke-NEHAR), v., to weaken, to debilitate. E»S Also WEAKENEAR, WEAKENIAR, WEAQUENEAR, WEA-QUENIAR, WIKENIAR, WIQUENEAR and WIQUE-NIAR.

wikeniar (wee·ke·NEHAR), v., to weaken, to debilitate. E»S Also WEAKENEAR, WEAKENIAR, WEAQUENEAR, WEA· QUENIAR, WIKENEAR, WIQUENEAR and WIQUE· NIAR.

wildear (weel·DEHAR), v., to go wild. Used also in the reflexive form. "Francisco se wildeó en la party." E»S Also WILDIAR.

wilderad (weel·the·RAD), n., f., wilderness. E»S

wildiar (weel·DEHAR), v., to go wild. Used also in the reflexive form. "Francisco se wildió en la party." E»S Also WILDEAR.

winchil (WIN·tchil), n., m., windshield. E»S Also GUINCHIL.

windo (WEEN·doh), n., f., computer window. "Close the windo so I can take out my diskette." {CS} E»S

windsurf (win·DSURF), n., m., water sport practiced on top of a board with a sail. E»S

windsurfista (win·sur·FIS·ta), n., m./f., a person who practices wind· surfing. E»S

winshi·waipers (ween·shee·WYE·pairs), n., m., windshield wipers. {CA} E»S

wiquenear (wee·ke·NEHAR), v., to weaken, to debilitate. E»S Also WEAKENEAR, WEAKENIAR, WEAQUENEAR, WEA· QUENIAR, WIKENEAR, WIKENIAR and WIQUE· NIAR.

wiqueniar (wee·ke·NEHAR), v., to weaken, to debilitate. E»S Also WEAKENEAR, WEAKENIAR, WEAQUENEAR, WEA· QUENIAR, WIKENEAR, WIKENIAR, and WIQUE· NEAR.

wiskería (wee·ke·REE·ah), n., f., whiskey store. E»S

wiski (WEES·kee), n., m., whisky. E»S Also GUISQUI.

withstandear (with-STAN-dehr), v., to withstand. E»S

wokman (WAK-man), n., f., walkman. E»S

wonderbrá (won-deer-BRA), n., m., bra with underwire and padding. {S} E»S

wop (WOP), n., f., person without papers. Eng. *without papers.* E»S

worestación (wore-sta-TION), n., f., work station. {CS} E»S

worshop (wor-SHOP), n., m., workshop. E»S

wrapear (wrah-PEHAR), v., to wrap. E»S

wrestlin (wrehs-LEEN), n., m., wrestling. Athletic activity. E»S

Y

Y.U.C.A. (YOO‑ka), n., Young Urban Cuban American. {C, CA}

yac (yah‑K), n., m., yak. E»S

yachear (YA‑chear), v., to use a yacht. E»S

yaching (YA‑chin), adv., competition sport practiced with sails. E»S

yaestufas (YA‑stu‑fas), exp., It's done! It's over! "Yaestufas, it's all over between us." {SW}

yanitor (YA‑ni‑tor), n., m., janitor. E»S

yanqui (YAN‑ki), n., m/f., American.

yaqui (YAH‑kee), n., f., jacket. "Qué guapa estás con tu nueva yaqui. How nice you look in your new jacket." {CA, NY} E»S

yarda (YAR‑da), n., f., backyard, patio. "Vayan a jugar en la yarda. Go play in the yard." {NY} E»S

yate (YA‑te), n., m., yacht. E»S

yaz (YAZ), n., m., jazz. E»S

yeguäenoche (ya-gueh-NOO-tche), n., f., nightmare. {NY}

yela (YE-la), n., f., jelly. {SW} E»S

yeld (YEL), n., m., economic benefit. Eng. *yield.* "Salió ganando un yeld. He earned a yield." E»S

yelo (YEH-loh), n., f., after the gelatin brand name Jello. E»S

yesquera (YES-ke-ra), n., f., tinderbox. {SW}

yet (YET), n., m., jet. E»S Also JET.

yiffie (YEE-fee), n., m/f., young professional in the 1990s. "Ellos fueron yiffies en Nuyol después de gradrarse del colegio."

yinyerel (yeen-YE-rel), n., m., ginger ale drink, after the soda brand. E»S.

yip (YEEP), n., m., jeep. E»S

yirish (YI-rish), n., m., Yiddish. "Es judío y habla yirish." E»S

yoguear (YO-gue-AR), v., to jog. E»S

yonko (YON-ko), n., m/f., 1. drug addict. 2. junkie. Also *yonqui.* "Es un pobre yonko. He's a miserable addict." {Mex, T}

yoquey (YO-key), n. m., 1. jockey. 2. racehorse rider. E»S Also YOQUI.

yoqui (YO-kee), n., m., jockey. E»S Also YOQUEY.

yorkshir (YOR-kshir), a., m., Yorkshire, type of dog. {Mex} E»S

yufo (YU-fo), n., Unidentified Flying Object. Also UFO E»S

yungla (YUN-glah), n., f., jungle. E»S.

yunier (JOO-NYOR) n., m., 1. wealthy young man. 2. Young person with political and financial connections. From Latin *juvenior.* "Ese yunier todavia tiene que aprender mucho." E»S Also JU-NIOR and YUNIOR.

yunior (JOO-NYOR) n., m., 1. wealthy young man. 2. Young person with political and financial connections. From Latin *juvenior.* "Ese yunior todavía tiene que aprender mucho." E»S Also JU-NIOR and YUNIER.

Z

zafacón (za‑pha‑KON), n., m., trash can. Eng. *safety can.* {D, NY, PR}
E»S Also SAFOCON.

zafo (ZA‑pho), exp., May the Lord protect me! "Con Zafo." {SW}
Also SAFO.

zai (ZAY), n., m., size, dimension, measurement. {CA} E»S

zap (ZAP), v., to change. E»S

zapeta (za‑PE‑ta), n., f, diaper. {Ch, SW}

zapo (ZA‑po), n., m., shoe. {Ch, SW}

zappin (ZAH‑pin), adv., change of TV channels through a remote con‑
trol.

zarzuela (zar‑SUE‑lah), n., f., popular Spanish musical comedy. The
term dates from 1888. S»E

zeta (ZE‑ta), n., m., defender of the oppressed. After Oscar "Zeta"
Acosta. {ELA, SW}

zipeado (zee-PEAH-do), adj., zipped. E»S

zipear (ZEE-peahr), v., to zip. E»S

zoquete (zo-KE-teh), adj., m., stupid. {SW}

zorrear (ZO-reahr), v., 1. to skip. 2. to take time off on the job. Sp. *zorro* means fox. {SW}

zorrero (zoh-RRE-ro), n., m., skipper. {SW}

zum (ZOOM), n., m., lens focalization. E»S

zumear (ZOO-mear), v., to zoom. E»S

Appendix

DON QUIXOTE DE LA MANCHA
MIGUEL DE CERVANTES

Transladado al Spanglish por Ilan Stavans

Nota Bene: The following exercise in literary translation came as a result of a fortuitous circumstance. During a lecture tour through Orwell's Catalonia in the summer of 2002, I participated in a radio discussion, broadcasted live, on the origins and nature of Spanglish. Among the participants, who were either present or connected by satellite, there was a language purist affiliated to the Real Academia Española de la Lengua Castellana. There was some discussion on the capacity of a language to express emotions and the challenge Spanglish faced in this area. At one point in a diatribe against Spanglish, this *caballero* stated that the mongrel tongue should not be taken seriously until and unless it produced a masterpiece of the caliber of *Don Quixote de La Mancha*, the magnum opus of Iberian letters, published by Cervantes in two parts, the first in 1605 and the second one in 1615. My immediate response was one I've repeated in the later section of "*La jerga loca*," the essay that serves as an introduction to this volume. It's too early to say what pattern Spanglish will take in its development, I suggested. While it isn't impossible that in a couple of hundred years such a masterpiece might be composed in a variety of Spanglish unfamiliar to us today, a "translation" of *Don Quixote* into Spanglish isn't at all impossible, and neither is it improbable.

Ipso facto, the program host asked me to improvise a few sentences. How would such translation "feel"? I spontaneously complied to his request—con enorme placer.

Upon my return to the downtown hotel, I spoke with Sergio Vila-Sanjuan, an editor for the daily *La Vanguardia*. After some discussion with his colaboradores y colegas, he wanted me to send him a.s.a.p. my translation of Part I, Chapter 1.

Poco después, at home already in Massachusetts, me dediqué de lleno to the endeavor. The piece appeared in the supplement *Cultura/s*.

An international controversy ensued.

The strategy I took to render the text is easy to summarize. Spanglish, as I've stated it in the meditative essay in this volume, remains, for the most part, an oral vehicle of communication, spoken predominantly by individuals of different national backgrounds in the United States. Although there is much in common among these national groups, each has devised its own linguistic modality. I refused to choose a single modality; instead, my objective, similar to those on Spanish-language TV north of the Rio Grande and, more important even, to the one used assiduously in the Internet, is a middle ground—de ningún lugar y de todas partes.

Mine isn't a standardized Spanglish because for now no such composite exists. Maybe soon, but not yet . . .

So mine is an "artificial" language, isn't it? Sure, it ought to be. Until and unless Spanglish moves from the oral to the written mode— and it's showing signs of doing so already—any literary attempt is, inevitably, una afectación. As translator I let myself be permeated by any and all varieties of Spanglish (Pachuco, Dominicanish, Cubonics, Nuyorrican, etc.) in the hope of producing a version that might be "read" by Latinos of different national backgrounds and by non-Hispanics and non-Spanglish-speakers as well.

A journalist for *La Nación* in Buenos Aires, not without sarcasm, described the effort as joyceano.

In Spanish, *joyceano* means Joycean; in Spanglish, joyful.

First Parte, Chapter Uno

In un placete de La Mancha of which nombre no quiero remembrearme, vivía, not so long ago, uno de esos gentlemen who always tienen una lanza in the rack, una buckler antigua, a skinny caballo y un greyhound para el chase. A cazuela with más beef than mutón, carne choppeada para la dinner, un omelet pa' los Sábados, lentil pa' los Viernes, y algún pigeon como delicacy especial pa' los Domingos, consumían tres cuarers de su income. El resto lo empleaba en una coat de broadcloth y en soketes de velvetín pa' los holidays, with sus slippers pa' combinar, while los otros días de la semana él cut a figura de los más finos cloths. Livin with él eran una housekeeper en sus forties, una sobrina not yet twenty y un ladino del field y la marketa que le saddleaba el caballo al gentleman y wieldeaba un hookete pa' podear. El gentleman andaba por allí por los fifty. Era de complexión robusta pero un poco fresco en los bones y una cara leaneada y gaunteada. La gente sabía that él era un early riser y que gustaba mucho huntear. La gente say que su apellido was Quijada or Quesada—hay diferencia de opinión entre aquellos que han escrito sobre el sujeto—but acordando with las muchas conjecturas se entiende que era really Quejada. But all this no tiene mucha importancia pa' nuestro cuento, providiendo que al cuentarlo no nos separemos pa' nada de las verdá.

It is known, pues, que el aformencionado gentleman, cuando se la pasaba bien, which era casi todo el año, tenía el hábito de leer libros de chivaldría with tanta pleasura y devoción as to leadearlo casi por completo a forgetear su vida de hunter y la administración de su estate. Tan great era su curiosidad e infatuación en este regarde que él even vendió muchos acres de tierra sembrable pa' comprar y leer los libros que amaba y carreaba a su casa as many as él podía obtuvir. Of todos los que devoreó, ninguno le plaseó más que los compuestos por el famoso Feliciano de Silva, who tenía un estylo lúcido y plotes intrincados that were tan preciados para él as pearlas; especialmente cuando readeaba esos cuentos de amor y challenges amorosos that se foundean por muchos placetes, por example un passage como this one: *La rasón de mi unrasón que aflicta mi rasón, en such a manera weakenea mi rasón que yo with rasón lamento tu beauty.* Y se sintió similarmente aflicteado cuando sus ojos cayeron en líneas como these ones: . . . *el high Heaven de tu divinidad te fortifiquea with las estrellas y te rendea worthy de ese deserveo que tu greatness deserva.*

El pobre felo se la paseaba awakeado en las noches en un eforte de desentrañar el meanin y make sense de pasajes como these ones, aunque Aristotle himself, even if él had been resurrecteado pa'l propósito, no los understeaba tampoco. El gentleman no estaba tranquilo en su mente por las wounds que dio y recebió Don Belianís; porque in spite of how great los doctores que lo trataron, el pobre felo must have been dejado with su face y su cuerpo entero coverteados de marcas y escars. Pero daba thanks al autor por concluir el libro with la promisa de una interminable adventura to come. Many times pensaba seizear la pluma y literalmente finishear el cuento como had been prometeado, y undoubtedly él would have done it, y would have succedeado muy bien si sus pensamientos no would have been ocupados with estorbos. El felo habló d'esto muchas veces with el cura, who era un hombre educado, graduado de Siguenza. Sostenía largas discusiones as to quién tenía el mejor caballero, Palmerín of England o Amadís of Gaul; pero Master Nicholas, el barbero del same pueblo, tenía el hábito de decir que nadie could come close ni cerca to the Caballero of Phoebus, y que si alguien *could* compararse with él, it had to be Don Galaor, bró de Amadís of Gaul, for Galaor estaba redy pa' todo y no era uno d'esos caballeros second-rate, y en su valor él no lagueaba demasiado atrás.

En short, nuestro gentleman quedó tan inmerso en su readin that él pasó largas noches—del sondáu y sonóp—y largos días—del daun al dosk—husmeando en sus libros. Finalmente, de tan poquito sleep y tanto readin, su brain se draidió y quedó fuera de su mente. Había llenado su imaginación con everythin que había readiado, with enchantamientos, encounters de caballero, battles, desafíos, wounds, with cuentos de amor y de tormentos, y with all sorts of impossible things, that as a result se convenció que todos los happenins ficcionales que imagineaba eran trú y that eran más reales pa' él que anithin else en el mundo. El remarcaba que el Cid Ruy Díaz era un caballero very good, pero que no había comparación with el Caballero de la Flaming Sword, who with una estocada had cortado en halfo dos giants fierces y monstruosos. El prefería a Bernardo del Carpio, who en Rocesvalles había slaineado a Roland, despait el charm del latter one, takin advantage del estylo que Hercules utilizó pa' strangulear en sus arms a Antaeus, hijo de la Tierra. También tenía mucho good pa' decir de Morgante, who, though era parte de la raza de giants, in which all son soberbios y de mala disposición, él was afable y well educado. But, encima de todo, él se cherisheaba de admiración por Rinaldo of Montalbán, especialmente when él saw him sallyingueando hacia fuera of su castillo pa' robear a todos los que le aparecían en su path, or when lo imagineaba overseas thifeando la statue de Mohammed, which, así dice la story, era all de oro. Y él would have enjoyado un mano-a-mano with el traitor Galalón, un privilegio for which él would have dado a su housekeeper y su sobrina en el same bargain.

In efecto, cuando sus wits quedaron sin reparo, él concebió la idea más extraña ever occurrida a un loco en este mundo. Pa' ganar más honor pa' himself y pa' su country al same time, le parecía fittin y necesario convertirse en un caballero errant y romear el mundo a caballo, en un suit de armadura. El would salir en quest de adventuras, pa' poner en práctica all that él readeaba en los libros. Arranglaría todo wrong, placeándose en situaciones of the greatest peril, and these mantendrían pa' siempre su nombre en la memoria. Como rewarda por su valor y el might de su brazo, el pobre felo podía verse crowneado por lo menos as Emperador de Trebizond; y pues, carriado por el extraño pleacer que él foundió en estos thoughts, inmediatamente he set to put el plan en marcha.

Lo primero que hizo fue burnishear old piezas de armadura, left to him por su great-grandfather, que por ages were arrumbada en una esquina, with polvo y olvido. Los polisheó y ajustó as best él could, y luego vio que faltaba una cosa bien importante: él had no real closed hemleto, but un morión o helmete de metal, del type que usaban los sol- dados. Su ingenuidad allowed him un remdio al bendear un cardbord en forma de half-helmete, which, cuando lo attacheó, dió la impresión de un helmete entero. Trú, cuando fue a ver si era strong as to withstandear un good slashin blow, quedó desappointeado; porque cuando dribleó su sword y dió un cople of golpes, succedió only en perder una semana entera de labor. Lo fácil with which lo había destrozado lo disturbó y de- cidió hacerlo over. This time puso strips de iron adentro y luego, con- vencido de que alredy era muy strong, refraineó ponerló a test otra vez. Instead, lo adoptó then y there como el finest helmete ever.

Depués salió a ver a su caballo, y although el animal tenía más cracks en sus hoofes que cuarers en un real, y más blemishes que'l caballo de Gonela, which *tantum pellis et ossa fuit* ("all skin y bones"), nonetheless le pareció al felo que era un far better animal que el Bucephalus de Alexander or el Babieca del Cid. El spent cuatro días complete tratando de encontrar un nombre apropriado pa'l caballo; porque—so se dijo to himself—viendo que era propriedad de tan famoso y worthy caballero, there was no rasón que no tuviera un nombre de equal renombre. El type de nombre que quería was one that would at once indicar what caballo it had been antes de ser propriedad del caballero errant y también what era su status presente; porque, cuando la condición del gentleman cambiara, su caballo also ought to have una apelación famosa, una high-soundin

one suited al nuevo orden de cosas y a la new profesión that was to follow; y thus, pensó muchos nombres en su memoria y en su imaginación
discardeó many other, añadiendo y sustrayendo de la lista. Finalmente
hinteó el de *Rocinante*, un nombre that lo impresionó as being sonoroso
y al same time indicativo of what el caballo had been cuando era de se-
gunda, whereas ahora no era otra cosa que el first y foremost de los ca-
ballos del mundo.

Habiendo foundeado un nombre tan pleasin pa' su caballo, decidió to
do the same pa' himself. Esto requirió otra semana. Pa'l final de ese
período se había echo a la mente that él as henceforth *Don Quixote*,
which, como has been stated antes, forwardeó a los autores d'este trú
cuento a asumir que se llamaba Quijada y no Quesada, as otros would
have it. Pero remembreando que el valiant Amadís no era happy que lo
llamaran así y nothin más, but addirió el nombre de su kingdom y su
country pa'cerlos famous también, y thus se llamó Amadís of Gaul; so
nuestro good caballero seleccionó poner su placete de origen y became
Don Quixote de La Mancha; for d'esta manera dejaría very plain su linaje
y confería honor a su country by takin su nombre y el suyo en one alone.

Y so, with sus weapons alredy limpias y su morión in shape, with
apelaciones al caballo y a himself, él naturalmente encontró que una sola
cosa laqueaba: él must seekiar una lady of whom él could enamorarse;
porque un caballero errant sin una ladylove was like un árbol sin leaves
ni frutas, un cuerpo sin soul.

"If," dijo, "como castigo a mis sines or un stroque de fortuna, me en-
cuentro with un giant, which es una thing que les pasa comunmente a
los caballeros errant, y si lo slaineo en un mano-a-mano o lo corto en
two, or, finalmente, si vanquisheo y se rinde, would it not be well tener
a alguien a whom yo puedo enviárselo como un presente, in order pa'
que'l giant, if él is livin todavía, may come in pa' arrodillarse frente a mi
sweet lady, y say en tono humilde y sumisivo, 'Yo, lady, soy el giant
Caraculiambro, lord de la island Malindrania, who has been derroteado

en un single combate por ese caballero who never can be praiseado enough, Don Quixote de La Mancha, el same que me sendió a presentarme before su Gracia pa' que Usté disponga as you wish?' "

Oh, cómo se revolotió en este espich nuestro good gentleman, y más than nunca él pensaba en el nombre that él should oferear a su lady! Como dice el cuento, there was una very good-lookin jovencita de rancho who vivía cerca, with whom él had been enamorado una vez, although ella never se dio por enterada. Su nombre era Aldonza Lorenzo y decidió that it was ella the one que debía to have el título de lady de sus pensamientos. Wisheó pa' ella un nombre tan good como his own y que conveyera la sugestión que era princeza or great lady; y, entonces, resolvió llamarla *Dulcinea del Toboso,* porque ella era nativa d'ese placete. El nombre era musical to his oídos, fuera de lo ordinario y significante, like los otros que seleccionó pa' himself y sus things.

Bibliography

This list isn't meant to be exhaustive. I've included the sources I used in the compilation of this volume as well as significant studies in lexicography, especially in dialectology. Also featured are titles on Ebonics, Yiddish, and other ethnic languages and dialects, as well as Spanish dictionaries of Anglicisms and essays on English loanwords from the Spanish language. And I've also inserted influential novels, collections of stories and volumes of poems whose content is either fully or partially en espanglés.

Acosta-Belén, Edna. "Spanglish: A Case of Languages in Contact," in *New Directions in Second Language Learning, Teaching and Bilingual Education*, M. Burt and H. Dulay, eds. Washington, DC: TESOL, 1975, 151–8.

Adams, Karen L., with Daniel T. Brink, eds. *Perspective on Official English: The Campaign for English as the Official Language of the U.S.A.* Berlin and New York: Mouton de Gruyer, 1990.

Adams, Ramón F. *Western Words: A Dictionary of the American West.* Norman, OK: University of Oklahoma Press, 1968.

Alemán, Mateo. *Ortografía castellana.* Mexico: El Colegio de México, 1950.

Alfaro, Ricardo J. *Diccionario de anglicismos.* Madrid: Editorial Gredos, 1964.

Algeo, John. "Spanish Loanwords in English by 1900," in *Spanish Loanwords in the English Language: A Tendency Towards Hegemony Reversal,* Félix Rodríguez González, ed., Berlin and New York: Mouton de Gruyter, 1996: 13–40.

Alonso Pedraz, Martín. *Evolución sintáctica del español.* Madrid: Aguilar, 1964.

———. *Gramática del español contemporáneo.* Madrid: Guadarrama, 1968.

Alzugara, J.J. *Diccionario de extranjerismos.* Madrid: Dossat, 1985.

Amastae, J., with L. Elias-Olivares, eds. *Spanish in the United States: Sociolinguistic Aspects.* Cambridge, Eng.: Cambridge University Press. 1982.

Atwood, E. Bagby. *The Regional Vocabulary of Texas.* Austin, TX: University of Texas Press, 1962.

Austin, Mary. "New Mexican Spanish," *Saturday Review of Literature* 7 (1931): 930.

Barker, George Carpenter. *Pachuco: An American Spanish Argot and Its Social Functions in Tucson, Arizona.* Tucson: University of Arizona Press, 1970.

Barreto, Amílcar A. *Language, Elites, and the State. Nationalism in Puerto Rico and Quebec.* Wesport, CT, and London: Praeger, 1998.

Bartlet, John Russell. *Dictionary of Americanisms. A Glossary of Words and Phrases Usually Regarded as Peculiar to the United States.* New York: Bartlett and Welford, 1848; 2nd ed, 1859.

Bello, Andrés, with Rufino José Cuervo. *Gramática de la lengua española.* Buenos Aires: Sopena, 1952.

Bentley, Harold W. *A Dictionary of Spanish Terms in English.* New York: Columbia University Press, 1932.

Bourke, John Gregory. "Notes on the Language and Folk Usage of the Rio Grande Valley," *Journal of American Folklore* 9 (1896): 81–115.

Brady, Haldeen. "Narcotic Argot Along the Mexican Border," *American Speech* 30 (1955): 84–90.

———. "Smugglers' Argot in the Southwest," *American Speech* 31 (1956): 96–101.

———. "The Pachucos and Their Argot," *Southern Folklore Quarterly* 24 (1960): 255–271.

Braschi, Giannina. *Yo-Yo Boing!* Pittsburgh, PA: Latin American Literary Review Press, 1998.

Burciaga, José Antonio. *Drink Cultura: Chicanismo.* Santa Barbara, CA: Joshua Odell Editions, 1993.

Campa, Arthur León. *Sayings and Riddles in New Mexico.* Albuquerque, NM: University of New Mexico Bulletin, Language Series 6, no. 2, 1937.

———. *Spanish Folk Poetry in New Mexico.* Albuquerque, NM: University of New Mexico Press, 1946.

Canfield, Delos Lincoln. *Spanish Pronunciation in the Americas.* Chicago, IL: University of Chicago Press, 1981.

Capdevila, Arturo. *Babel y el castellano.* Buenos Aires: Editorial Losada, 1945.

———. *Despeñaderos del habla: Negligencia, cursilería, tuntún.* Buenos Aires: Editorial Losada, 1952.

Carreter, Fernando Lázaro. *Crónica del "Diccionario de Autoridades" (1713–1740).* Madrid: Eosgraf, S.A., 1972.

———. *El dardo en la palabra.* Madrid: Editorial Guttemberg-Círculo de Lectores, 1997.

Casiano Montañez, L. *La pronunciación de los puertorriqueños en Nueva York.* Bogotá: Ediciones Tercer Mundo, 1975.

Chabat, Carlos G. *Diccionario de caló. El lenguaje del hampa en México.* Mexico: Librería de Medicina, 1964.

Chávez Silverman, Susana. *Killer Crónicas.* Madison, WI: University of Wisconsin Press, 2004.

Cisneros, Sandra. *Caramelo.* New York: Random House, 2002.

Cobos, Rubén. "The New Mexican Game of Valse Chiquiao," *Western Folklore* 15, no. 2 (1956).

———. *Refranes españoles del sudoeste. Southwest Spanish Proverbs.* Cerrillos, NM: San Marcos Press, 1974.

———. *A Dictionary of New Mexico and Southern Colorado Spanish.* Santa Fe, NM: Museum of New Mexico Press, 1983.

Coll y Toste, Cayetano. "Vocabulario de palabras introducidas al idioma español, procedentes del lenguaje indo-antillano," *Boletín Histórico de Puerto Rico,* viii (1921): 294.

Conklin, Nancy Faires, with Margaret A. Lourie. *A Host of Tongues: Language Communities in the United States.* New York: Free Press, 1983.

Corominas, Joan, with José A. Pascual. *Diccionario crítico etimológico castellano e hispánico.* Madrid: Gredos, 1980.

Covarrubias Orozco, Sebastián de. *Tesoro de la lengua castellana o española.* Edited by Felipe C.R. Maldonado, revised by Manuel Camarero. 2nd edition corrected. Madrid: Editorial Castalia, 1995.

Crawford, James. *Bilingual Education: History, Politics, Theory, and Practice.* Trenton, NJ: Crane Publishing, 1989.

———. *Hold Your Tongue: Bilingualism and the Politics of "English-Only."* Reading, MA: Addison-Wesley, 1992.

Criado de Val, Manuel. *Diccionario del español equívoco.* Madrid: Sgel, S.A., 1981.

Cruz, Bill, with Bill Teck, et al. *The Official Spanglish Dictionary.* New York: Fireside, 1998.

Cuyás, Arturo. *Appleton's New Cuyás English-Spanish, Spanish-English Dictioinary.* Englewood Cliffs, NJ: Prentice-Hall, 1972.

De Miguel, Amando. *La perversión del lenguaje.* Barcelona: Espasa Calpe, 1985.

Díaz, Junot. *Drown.* New York: Riverhead, 1996.

Diccionario de Autoridades de la Real Academia Española. 2 volumes. Madrid: Editorial Gredos, 1990.

Diccionario de la lengua española de la Real Academia. 22nd edition. Madrid: Espasa Calpe, 2001.

Diccionario del español usual en México. Mexico City: Colegio de México, 1996.

Dillard, Joey Lee. *Black English: Its History and Usage in the United States.* New York: Vintage, 1972.

———. *Toward a Social History of American English.* New York: Mouton, 1985.

———. *A History of American English.* New York: Longman, 1992.

Doval, Gregorio. *Diccionario de expresiones extranjeras.* Madrid: Ediciones del Prado, 1996.

Elías-Olivares, Lucía, ed. *Spanish Language Use and Public Life in the United States.* Berlin and New York: Mounton, 1985.

Elliott, John Huxtable. *Imperial Spain: 1469–1716.* London: Edward Arnold, 1963.

Espinosa, Aurelio M. *The Spanish Language in New Mexico and Southern Colorado.* Historical Society of New Mexico, Publication 16. Santa Fe, NM, 1911.

Fought, C. *The English and Spanish of Young Adult Chicanos.* Doctoral Thesis. University of Pennsylvania, 1997.

Fuentes, Dagoberto, with José A. López. *Barrio Language Dictionary: First Dictionary of Caló.* Los Angeles, CA: Southland Press/El Barrio Publications, 1974.

Galván, Roberto A., with Richard V. Teschner. *Diccionario del español chicano.* Lincolnwood, IL: National Textbook Company, 1989.

Gamio, Manuel. *Mexican Immigration to the United States: A Study of Human Migration and Adjustment.* Chicago, IL: University of Chicago Press, 1930.

———. *The Life Story of the Mexican Immigrant: Autobiographical Documents.* Introduction by Paul S. Tayler. New York: Dover Publications, 1971.

Gómez-Peña, Guillermo. *Warrior for Gringostroika: Essays, Performance Texts and Poetry.* St. Paul, MI: Graywolf Press, 1993.

———. *The New World Border: Prophecies, Poems and Loqueras for the End of the Century.* San Francisco: City Lights, 1996.

———. *Dangerous Border Crossers: The Artist Talks Back.* London: New York: Routledge, 2000.

———, with Enrique Chagoya and Felicia Rice. *Codex Espangliensis: From Columbus to the Border Patrol.* San Francisco: City Lights Books, 2000.

Gray, Edward D. McQueen. *Spanish Language in New Mexico: A National Resource.* University of New Mexico Bulletin, Sociological Series 1, no.2, 1912.

Grose, Francis. *A Classical Dictionary of the Vulgar Tongue.* Edited by Eric Partridge. London: Scholartis Press, 1931.

Grosjean, François. *Life with Two Languages: An Introduction to Bilingualism.* Cambridge, MA: Harvard University Press, 1982.

Grosschmid, Pablo, with Cristina Echegoyen. *Diccionario de regionalismos de la lengua española.* Barcelona: Editorial Juventud, 1997.

Gutiérrez, Félix F., with Jorge Reina Schement. *Spanish-Language Radio in the Southwestern United States.* Austin, TX: University of Texas Press-Center for Mexican-American Studies, 1979.

Haensch, Günther. *Los diccionarios del español en el umbral del siglo XXI.* Salamanca, Spain: Ediciones Universidad de Salamanca, 1997.

Hakuta, Kenji. *Mirror of Language: The Debate on Bilingualism.* New York: Basic Books, 1986.

Hamers, Josiane F. with Michael H. A. Blanc. *Bilinguality and Bilingualism.* Cambridge, Eng.: Cambridge University Press, 1989.

Haugen, Einar Ingvald. *Bilingualism in the Americas: A Bibliography and Research Guide.* Gainesville, FL: American Dialect Society; University, AL: University of Alabama Press, 1956.

Henríquez-Ureña, Pedro. "Observaciones sobre el español de América," *Revista de Filología Española,* viii (1921) and xvii (1930).

———. *El español en Santo Domingo.* Santo Domingo, R.D.: Taller, 1975. First published in 1940.

Hernández-Chávez, F., with A. Cohen and A. Beltramo, eds. *El lenguaje de los Chicanos: Regional and Social Characteristics of Language Used by Mexican Americans.* Arlington, VA: Center or Applied Linguistics, 1975.

Herrera, Juan Felipe. *Border-Crosser with a Lamborghini Dream.* Tucson, AZ: University of Arizona Press, 1999.

———. *CrashBoomLove: A Novel in Verse.* Albuquerque, NM: University of New Mexico Press, 1999.

Hoffman, Charlotte. *An Introduction to Bilingualism.* New York: Longman, 1991.

Irwin, Godfrey. *American Tramp and Underworld Slang: Words and Phrases Used by Hoboes, Tramps, Migratory Workers and Those on the Fringes of Society, with Their Uses and Origins, with a Number of Tramp Songs.* London: Scholartis Press, 1931.

Johnson, Fern J. *Speaking Culturally: Language Diversity in the United States.* Thousand Oaks, CA: Sage Publications, 2000.

Johnson, Samuel. *A Dictionary of the English Language.* London, 1755. See also *Johnson's Dictionary: A Modern Selection,* edited by E. L. McAdam, Jr. and George Milne. New York: Pantheon, 1964.

Kiddle, Lawrence B. "Los nombres del pavo en el dialecto nuevomejicano." *Hispania* 24 (1941): 213–16.

———. " 'Turkey' in New Mexican Spanish." *Romance Philology* 5 (1951–52): 190–97.

Klee, Carol A., with Luis Ramos García, eds. *Sociolinguistics of the Spanish-speaking World: Iberia, Latin America, the United States.* Tempe, Arizona: Bilingual Press, 1991.

Kreidler, Charles W. *A Study of the Influence of English on the Spanish of Puerto Ricans in Jersey City, New Jersey.* Doctoral Thesis. University of Michigan, 1957.

Labov, William. *A Study of Non-Standard English of Negro and Puerto Rican Speakers in New York City.* New York: Columbia University Press, 1968.

Laviera, Tato. *La Carreta Made a U-Turn.* Houston, TX: Arte Público Press, 1979.

——. *Enclave.* Houston, TX: Arte Público Press, 1981.

——. *AmeRícan.* Houston, TX: Arte Público Press, 1985.

——. *Mainstream Ethics = Etica corriente.* Houston, TX: Arte Publico Press, 1988.

Lawton, D. "Chicano Spanish: Some Socioeconomic Considerations," *Bilingual Review* 2,3 (1976): 22–33.

León, Aurelio de. *Barbarismos comunes en México.* 2 vols. Mexico: Imprenta mundial, 1936, 1937.

León-Portilla, Miguel, with Earl Shorris, et al. *In the Language of Kings: An Anthology of Mesoamerican Literature.* New York: W.W. Norton, 2001.

Lighter, Jonathan L., ed. *Random House Historical Diccionary of American Slang.* Vols. 1 to 3. J. Ball and J. O'Connor, assistant eds. New York: Random House, 1994.

Lipski, John M. *Linguistic Aspects of Spanish-English Language Switching.* Tempe, AZ: Arizona State University, Center for Latin American Studies, 1985.

——. *The Language of the 'Isleños.' Vestigial Spanish in Louisiana.* Baton Rouge, LA, and London: Louisiana University Press, 1990.

——. *Latin American Spanish.* New York: Longman, 1994.

Lockhart, James. *Nahuas and Spaniards: Postconquest Central Mexican History and Philology.* Stanford, CA: Stanford University Press; and Los Angeles, CA: Latin American Center Publications, University of California, 1991.

Lope Blanch, Juan M. *El español hablado en el sureste de los Estados Unidos. Materiales para su estudio.* Mexico: Universidad Nacional Autónoma de México, 1990.

Maitland, James. *American Slang Dictionary*. Chicago, IL. 1891.

Major, Clarence. *Juba to Jive: A Dictionary of African American Slang*. New York: Viking, 1994.

Malaret, Augusto. *Diccionario de provincialismos de Puerto Rico*. San Juan, PR: 1917.

———. *Diccionario de americanismos*. Buenos Aires: Academia Argentina de las Letras, 1942.

Maldona González, Concepción, ed. *Clave: Diccionario del uso del español actual*. Prologue by Gabriel García Márquez. Madrid: Ediciones SM, 1996.

Martínez de Sousa, José. *Dudas y errores del lenguaje*. Madrid: Editorial Paraninfo, 1992.

McAdam, Edward Lippincott, Jr., with George Milne. *Johnson's Dictionary: A Modern Selection*. New York: Pantheon Books, 1963.

McMahon, April M.S. *Understanding Language Change*. Cambridge, Eng.: Cambridge University Press, 1994.

Mencken, Henry Louis. *The American Language*. New York: Alfred A. Knopf, 1980.

Menéndez Pidal, Ramón. *Manual de gramática histórica española*. Madrid: Espasa-Calpe, 1949.

———. *Orígenes del español*. Madrid: Espasa-Calpe, 1950.

Milroy, Lesley, with Pieter Muysken, eds. *One Speaker, Two Languages: Cross-disciplinary Perspectives on Code-Switching*. Cambridge, Eng.: Cambridge University Press.

Mir, Juan. *Rebusco de voces castizas*. Madrid: 1907.

Moliner, María. *Diccionario del uso del español*. 2 volumes. 2nd edition. Madrid: Editorial Gredos, 1998.

Motolinía, Toribio de. *History of the Indians of New Spain*. Translated and edited by Elizabeth Andras Foster. Berkeley, CA: Cortés Society, 1950.

Mufwene, Salikoko S., with R. Rickford, G. Bailey, and J. Baugh, eds. *African American English: Structure, History, and Use*. New York: Routledge, 1998.

Muñoz Cortés, Manuel. *El español vulgar*. Madrid, 1959.

Myers-Scotton, Carol. *Social Motivations for Codeswitching: Evidence from Africa*. Oxford, Eng.: Clarendon Press, 1993.

Nash, R. "Spanglish: Language Contact in Puerto Rico," *American Speech* 45 (1970): 223–33.

Nebrija, Antonio de. *Gramática de la lengua castellana* (Salamanca, 1492). Along with *Muestra de la istoria de las antiguedades de España and Regla de orthographia de la lengua castellana.* Edited, with an introduction and notes, by Ignacio González-Llubera. Oxford and London: Oxford University Press, 1926.

Neuman, Henry, with Giuseppe Marco Baretti. *Dictionary of the Spanish and English Languages.* 2 vols. Boston, MA: Walkins, Carter, and Co., 1847.

Ornstein-Galicia, Jacob. ed. *Form and Function of Chicano English.* Rowley, MA: Newbury House, 1988.

Ortego, P.D. "Some Cultural Implication of a Mexican-American Border Dialect of American English," *Studies in Linguistics* 21 (1970): 77–82.

Ortiz, Fernando. *Un catauro de cubanismos.* Havana, 1923.

———. *Glosario de afronegrismos.* Havana: Imprenta 'El Siglo XX,' 1924.

———. *Nuevo catauro de cubanismos.* Havana: Editorial de Ciencias Sociales, 1974.

Oxford English Dictionary. Oxford, Eng., and New York: Oxford University Press, 1995.

Padilla, Francisco. *Bilingual Dictionary of Anglicismos, Barbarismos, Pachuquismos y Otras Locuciones en El Barrio.* Denver, CO, 1980

Padilla, Raymond V., with Alfredo H. Benavides, eds. *Critical Perspectives on Bilingual Education Research.* Tempe, AZ: Bilingual Review Press, 1992.

Partridge, Eric, with John W. Clark. *British and American English since 1900, with Contributions on English in Canada, South Africa, Australia, New Zealand and India.* New York: Philosophical Library, 1951.

———. *Slang To-day and Yesterday, with a Short Historical Sketch and Vocabularies of English, American, and Australian Slang.* London: Routledge & Kegan Paul, 1950.

———. *Origins: A Short Etymological Dictionary of Modern English.* London: Routledge & Kegan Paul, 1958.

———. *A Charm of Words: Essays and Papers on Language.* London: Hamish Hamilton, 1960.

———. *The Gentle Art of Lexicography as Pursued and Experienced by an Addict.* New York: Macmillan, 1963.

————. *A Dictionary of Catch Phrases, British and American, from the Sixteenth Century to the Present Day*. Briarcliff Manor, NY: Stein and Day, 1977.

————. *A Dictionary of Slang and Unconventional English: Colloquialisms and Catch-Phrases, Solecisms and Catachreses, Nicknames, and Vulgarisms*. Edited by Paul Beale. New York: Macmillan, 1984.

————. *Usage and Abusage: A Guide to Good English*. Edited by Janet Whitcut. New York: W.W. Norton, 1995.

Paz, Octavio. *The Labyrinth of Solitude*. Translated by Lysander Kemp. New York: Grove, 1962.

Peñalosa, Fernando. *Chicano Sociolinguistics*. Rowley, MA: Newbery House, 1980.

Pérez, Bertha with María E. Torres-Guzmán. *Learning in Two Worlds: An Integrated Spanish/English Biliteracy Approach*. New York: Longman, 1992.

Peyton, Elizabeth V. with Guillermo Rojas Carrasco. *Anglicismos*. Valparaíso, Chile: Editorial Amanecer, 1944.

Poplack, Shana. "Sometimes I'll Start a Sentence in Spanish *y termino en español:* Toward a Typology of Code-Switching," *Linguistics* 18 (1980): 581–616.

Rael, Juan B. " 'Cosa nada' en el español nuevomejicano." *Modern Language Notes* 69, no. 1 (1934): 31–32.

Ramos, Jorge. *The Other Face of America: Chronicles of the Immigrants Shaping Our Future*. Translated by Patricia J. Duncan. New York: Rayo/HarperCollins, 2002.

Ramos Duarte, Félix. *Diccionario de mexicanismos*. Mexico, 1898.

Richard, Renaud, ed. *Diccionario de hispanoamericanismos no recogidos por la Real Academia*. Madrid: Cátedra, 1997.

Roca, Ana, with John B. Jensen, eds. *Spanish in Contact: Issues in Bilingualism*. Somerville, MA: Cascadilla Press, 1996.

————, with John M. Lipski. *Spanish in the United States: Linguistic Contact and Diversity*. New York: Mouton de Gruyter, 1993.

Rodríguez, Richard. *Hunger of Memory: The Education of Richard Rodríguez*. Boston, MA: David R. Godine, 1982.

————. *Brown: The Last Discovery of America*. New York: Viking, 2002.

Rodríguez González, Félix, ed. *Spanish Loanwords in the English Language: A Ten-*

dency Towards Harmony Reversal. Berlin, Ger., and New York: Mouton de Gruyter, 1996.

———. *Nuevo diccionario de anglicismos.* Madrid: Gredos, 1997.

Romaine, Suzanne. *Bilingualism.* Oxford, Eng.: Basil Blackwell, 1989.

———, ed. *The Cambridge History of the English Language.* Cambridge, Eng.: Cambridge University Press, 1992, 1999.

Rosario, Rubén del. *La lengua de Puerto Rico: Ensayos.* 6th edition, revised. Río Piedras, PR: Editorial Cultural, 1969.

Rosenblat, Angel. *Buenas y malas palabras en el castellano de Venezuela.* Caracas and Madrid, 1956 and 1960.

———. *El castellano de España y el castellano de América. Unidad y diferenciación.* Caracas: Universidad Central de Caracas, 1962.

Rubio, Darío. *La anarquía del lenguaje en la América española.* Mexico, 1925.

Sánchez, Rosaura. *Chicano Discourse: Socio-Historic Perspectives.*

Sánchez-Boundy, José. *Diccionario de cubanismos más usuales.* Vols. I to VI. Miami: Ediciones Universal, 1989.

———. *Diccionario Mayor de Cubanismos.* Miami, FL: Ediciones Universal, 1999.

Santamaría, Francisco J. *Diccionario general de americanismos.* 3 vols. Mexico: Editorial Pedro Robredo, 1942.

———. *Diccionario de mejicanismos.* Mexico: Editorial Porrúa, 1959.

———. *Ensayos críticos del lenguage.* Mexico: Editorial Porrúa, 1980.

Seco, Manuel, with Olimpia Andrés and Gabimo Ramos. *Diccionario de dudas y dificultades de la lengua española.* Madrid: Espasa-Calpe, 1986.

———. *Diccionario del español actual.* 2 volumes. Madrid: Aguilar, 1999.

Serrano, Rodolfo G. *Dictionary of Pachuco Terms.* Bakersfield, CA: Sierra Printers, 1976.

Skutnabb-Kangas, Tove. *Bilingualism or Not: The Education of Minorities.* Clevedon, Avon, Eng.: Multilingual Matters, 1984.

Smitherman, Geneva. *Black Talk. Words and Phrases from the Hood to the Amen Corner.* Boston and New York: Houghton Mifflin, 1994.

Taschner, Richard V., with Garland D. Billis and Jerry R. Craddock, eds. *Spanish and English of the United States Hispanos: A Critical, Annotated, Linguistic Bibliography.* Arlington, VA: Center for Applied Linguistics, 1975.

Thomas, Piri. *Down These Mean Streets.* New York: Knopf, 1967.

Tomarón, Marqués de [Santiago de Mora-Figueroa y Williams], with Jaime Otero, eds. *El peso de la lengua española en el mundo.* Valladolid, Spain: Fundación Duque de Soria and INCIPE, 1995.

Turner, Paul R., ed. *Bilingualism in the Southwest.* Tucson, AZ: University of Arizona Press, 1973.

Vaquero, María. *Palabras son palabras.* San Juan, PR: Editorial Plaza Mayor.

Varela, Beatriz. *El español cubano-americano.* New York: Senda Nueva Ediciones, 1992.

Vasconcelos, José. *The Cosmic Race/La raza cósmica.* Translated by Didier T. Jaén. Baltimore, MD: The Johns Hopkins University Press, 1997.

Vázquez, Librado Keno, with María Enriqueta Vázquez, *Regional Dictionary of Chicano Slang.* Austin, TX: Jenkins Publishing Co./The Pemberton Press, 1975.

Vega, Ana Lydia, with Carmen Lugo Filippi. *Vírgenes y mártires.* Rio Piedras, PR: Antillana, 1981

Vega, Garcilaso de la. *Royal Commentaries of the Incas.* Translated by Harold V. Livermore. Austin, TX: University of Texas Press, 1966.

Veltman, Calvin. *The Future of the Spanish Language in the United States.* New York and Washington, DC: Hispanic Policy Development Project, 1983.

———. *Language Shift in the United States.* New York: Mouton, 1983

Weekley, Ernest. *The Romance of Words.* New York: Dover Publications, 1911.

Weinreich, Uriel. *Languages in Contact: Findings and Problems.* The Hague, Holland: Mouton, 1953.

Wilber, Cynthia J., with Susan Lister. *Medical Spanish: The Instant Survival Guide.* 3rd edition Boston: Butterworth-Heinemann, 1995.

Wolfram, Watt. *Sociolinguistic Aspects of Assimilation: Puerto Rican English in New York City.* Arlington, VA: Center for Applied Linguistics, 1974.

Zamora, Vicente A. *Dialectología española.* 2nd edition. Madrid: Gredos, 1979.

Zentella, Ana Celia. *Growing Up Bilingual: Puerto Rican Children in New York.* Malden, MA: Blackwell, 1997.

Acknowledgments

The roots—las rutas—of this book are labyrinthine. The list of recipients of my wholehearted gratitude starts with Charles M. Levine at Random House, whose idea it was to compile a dictionary of Spanglish. He suggested it to me in a luncheon in the Upper East Side. My recollection covers the event under a veil of falsity: Did Levine truly recognize Spanglish as a legitimate means of communication? Was he sarcastic when he challenged me to embark on the project? My friend Philip Lief had orchestrated the luncheon. He too doubts that a division as prestigious as the one devoted to reference books would endorse so controversial a lexicon. But he kept at my side, pushing me to complete the lexicon, providing me with advice and assistance. After an itinerant journey, René Alegría at Rayo/HarperCollins came to the rescue. The endeavor suited his vision of una América nueva. He signed the book and graciously pushed me. It has taken me more than half a decade to complete it. Happily, the years have gone by with Alegría (in Spanish, his last name means "happiness") as a friend at my side.

I owe much to literally hundreds of people—perhaps even thousands—with whom I held correspondence over the years. Although it felt as if I had them at my fingertips, they were in faraway places: all across the United States, from Seattle to UCLA, from Austin to Chicago, from Tallahassee to San Juan; everywhere in the Americas, from Buenos Aires, Santiago and Bogotá to Managua, Lima, Distrito Federal, and Havana; and also in Spain, Norway, Italy, France, Germany, and Israel. By definition, a list of their names would be incomplete.

My first published piece on the topic was "The Sounds of Spanglish," in *Hopscotch* 1,1 (1999). My appreciation to the editorial staff of the journal and to its publisher, Duke University Press. Mike Vazquez, managing editor at *Transition,* offered advise on the piece. I was then invited to lecture in Madrid's Casa de América. My speech was published in a booklet entitled *Spanglish para millones* (2000). Dozens of engagements took place successively at institutions such as Harvard University, University of London-ILAS, Instituto Cervantes, Cambridge University, M.I.T., University of Michigan, The Terra Museum in Chicago, University of California at Berkeley, and Universidad de Barcelona. My deep appreciation to the organizers of these events. Earlier drafts of the introduction in this book have appeared, in somewhat different form, in the *Chronicle of Higher Education, World Literature Today, The Cambridge Companion to Latin American Culture* edited by John King, and *Lives in Translation: Bilingual Writers on Identity and Creativity,* edited by Isabelle de Courtivron. I further developed my argument in my column published in *Cuadernos Cervantes,* under the guidance of David Hernández de la Fuente. And it sidetracked in pieces for *The Forward, The Nation, The Boston Globe, Lateral, The Miami Herald,* and *Foreign Affairs en Español.*

I found the best forum for debate to be the countless interviews I've granted to reporters, newscasters, and journalists. Their questions were incisive, pushing me to articulate my thoughts further. They also allowed me to understand the fluidity of a lexicon in its transitional stage from the oral to the written format. My special gratitude to Jorge Ramos of *Noticiero Univisión*, Don Francisco of *Sábado Gigante*, the staff of BBC radio, *The New York Times*, *L'Unità* in Italy, *Lateral* in Barcelona, Radio Netherlands, Televisa, *O Globo* in Brazil, as well as CNN and NPR.

Friends and colleagues either read versions of the manuscript or discussed the ideas in public forums. I appreciate the support of César Alegre, Julia Alvarez, Nuria Amat, Homero Aridjis and Betty Ferber, Ricardo Armijo, Marc Aronson, Harold Augenbraum, Ruth Behar, Antonio Benítez-Rojo, Alicia Borinsky, Giannina Brasci, Cass Canfield, Jr., Carmen M. Carracelas-Juncal, David Carrasco, Robin Cembalest, Robert Con Davis, Juan Cruz, Joie Davidow, Rosanne Dávila, Mihály Dés, Junot Díaz, Morris Dickstein, Ariel Dorfman, Martín Espada, Christian Faltis, Morris Farhi, Oscar H. Faúndez, Rosario Ferré, Rossana Fuentes Berain, Alberto Fuguet, Héctor García, Isaac Goldemberg, Francisco Goldman, Guillermo Gómez-Peña, Rigoberto González, Menene Grass, Roberto Guareschi, Patti Hartmann, Anne "Nancy" Hartzenbusch, Bobbie Helinski, David Hernández de la Fuente, Mónica Herrero, Rolando Hinojosa-Smith, María Herrera-Sobek, Roxana Kahale, Steven G. Kellman, John King, Lalo López Alcaraz, James Maraniss, Jaime Manrique, Roberto Márquez, Demetria Martínez, Angelina Muniz-Huberman, Ed Morales, Alvaro Mutis, Gustavo Pérez-Firmat, Danny Postel, Achy Obejas, Hilda Otaño-Benítez, Edmundo Paz-Soldán, Gregory Rabassa, Julia Reidhead, Félix Rodríguez, Luis J. Rodríguez, Richard Rodríguez, Félix Rodríguez González, Anthony Rudolf, Esmeralda Santiago, Donna Sanzonne, Moacyr Scliar, Roberto Schwartz, Stephen Sadow, John Philip Santos, José Saramago, Guillermo Schavelzon, Michelle Serros, Earl and Sylvia Shorris, Ana María Shua, Neal Sokol, Leo Spitzer, Eduardo Subirats, Doris Sommer, Sandy Taylor, Antonio Torres, Angela Torres-Henrick, Silvio Torres-Saillant, Sergio Troncoso, Ana Lydia Vega, Francisco

Vega, Irene Vilar, Sergio Vila-Sanjuan, Gerardo and Susana Villacrés, Teresa Villegas, Tino Villanueva, Karen Winkler, Art Winslow, Vicente A. Zamora, and Ana Celia Zentella. I sincerely appreciate their advice.

The producer Joseph Tovares at PBS-WGBH allowed me to further explore the theory behind this lexicon in the series *La Plaza: Conversations with Ilan Stavans*. He and his assistant Margaret Carsley became confidants. The counter-voices of Juan Flores and Roberto González-Echeverría were infinitely helpful to me.

My translation into Spanglish of *Don Quixote de La Mancha* Part I, Chapter 1, was first published by Eduardo Vila-Sanjuan in Barcelona's *La Vanguardia* (3 July 2002): 5–6.

My students at Amherst College in the course *The Sounds of Spanglish* compiled voices, tested ideas with courage and intelligence, and experimented with countless verbal transactions. I especially benefited from the work done by Nathan Clay, Danielle Billinkoff, Elisa Cantero, Marisol Corral, Joel Estrada, Lisa Garibay, Karen Grawjer, Melissa Lorenzo, Michael Pages, Adriana Mariño, Megan Richetti, Perla Roffe, Brian Sheehy, Nikki Tshuwaka, Max Ubelaker, Beatriz Verdasco-Vidal, and Salomón Zavala. Also, my gratitude to their relatives, who served as electronic and tape-recording correspondents: Raquel Aspuru, María Alina Lorenzo, Angel E. Lorenzo, Michelle T. Lorenzo, Michael S. Lorenzo, Andrea J. Espinosa, Flavia Lorenzo, Bertha Merzeau, Raúl Sanchez, and José A. Vázquez.

Thanks also to Don Fehr at Basic Books and Gayatri Patnaik at Routledge. Bobbie Helinski helped with the endless administrative aspects involved in the preparation of this lexicon. As always, my assistant Jennifer M. Acker was an invaluable resource in tasks too many to count. Andrea Montejo gracefully and smoothfully moved the manuscript along the editorial process. Juan Pablo Lombana and Robert Legault did remarkable jobs copy-editing and proofreading. Mil gracias.